Practical Projects

Edited by
George Brown, PhD, CEng, FIEE, M5ACN

Radio Society of Great Britain

Published by the Radio Society of Great Britain, Cranborne Road, Potters Bar, Herts EN6 3JE.

First published 2002.

ISBN 1 872309 88 7

Cover design: Braden Threadgold, Potters Bar.
Illustrations: Bob Ryan.
Subediting and typography: Ray Eckersley.
Production: Mark Allgar.

Printed in Great Britain by PWP Acrolith Printing Ltd, Hertford (www.pwpacro.com).

Contents

Preface

This book is the natural successor to the *Radio & Electronics Cookbook*, published some time ago. It proved very popular both with newcomers to the hobby and with 'old hands' who wanted to try something not too difficult with a technology that was not around in their formative years. It also provided everyone who read it with a source of ideas for other projects, which is what amateur radio is all about.

It is a pleasure to acknowledge the assistance of the New Zealand Association of Radio Transmitters (NZART) for its contribution of some articles. These originated in *Break-In*, the official NZART journal, and were provided by its Editor, John Walker, ZL3IB. Thank you, John.

There remains the ever-present problem of sourcing those components which disappear from the shelves of stockists between the submission of the articles and the appearance of the book. This cannot be avoided. Radio amateurs, over the years, have developed the instincts of the magpie, and can be seen at rallies poring over junk to see if that elusive chip or transistor can be found for their latest project. While being a valuable attribute, hoarding must be done in moderation, otherwise the rapid accumulation of junk will soon alienate the other members of your family. If that component still eludes you, why not ask members of your local radio club? They will be pleased to help. If you don't know where your nearest club is, a phone call to the RSGB will supply you with a contact name and telephone number.

This book differs slightly from the previous one in that it addresses the problem faced by all constructors at one time or another: "Now that I've built it, what do I do with it?" There is a reference section at the end of the book which gives some ideas, and also provides some information on specialised components and techniques, along with the much-neglected subject of safety in the shack.

Nevertheless, the major portion of the book is devoted to weekend projects. Choose your projects carefully according to your needs and experience, and I hope you will derive as much pleasure from their construction and use as I did from collecting them together.

George Brown
Potters Bar, 2002

A loop antenna for hearing aids

Not only have my eyes become dim over the last few years but also my ears! I have had to succumb to the assistance of a hearing aid and my life has been transformed. I can hear again and to a degree can appreciate music. Little compensation, perhaps, but it is not all gloom. Some hearing aids have a very useful built-in function, a loop reception system. This design is for those of us who use hearing aids. Loop systems are available commercially and use active microphones and amplifiers. This unit will cost less than a pound and will be a worthwhile addition to the shack.

Hearing loss is not always constant across the whole audio spectrum; mine is good to 500Hz, then drops off to –80dB by 2kHz. This makes the use of headphones difficult, even when we have mastered receiving Morse using a low-frequency tone. With the aid, my hearing is now reasonably flat (within 20dB!) from 200Hz to 3.5kHz, the problem now being that I cannot wear headphones with the aid.

THE LOOP

If we could tap into the loop system we would get reasonable reproduction of SSB and FM signals, and on Morse we could use all the receiver's functions to the full. I doubt if those of you with good hearing can imagine tuning a notch filter when you cannot hear the notch affecting the high frequencies; it makes tuning much sharper for us. Another advantage is a reduction in noise spill-out from the headphones, enabling us to listen without distracting others from their pleasures.

The hearing aid can be switched into loop mode, which disconnects the miniature internal microphone and connects the pick-up loop. If we now place the aid within a varying magnetic field, driven by the output of a receiver, we will hear the audio directly in the ear. This is a lot less difficult than first imagined. I did think that I would need an amplifier to drive the loop to give a sufficiently strong magnetic field, but I was wrong. My first test coil was a small drum of hook-up wire connected to the headphone socket of the receiver and I could hear up to six feet away. The main problem with this was that head orientation was important to get the best pick-up.

The next test coil was wound round the top of an ice cream container and gave a useful level of listening up to a foot away. A loop this size hung round the neck is ideal and is the basis of the expensive commercial units. The next one used a loop 14in square and this gave good levels when lying on the bench, giving completely free movement. In fact I am using it to listen to the radio while typing this. The third and final test loop was wound round the room from floor to ceiling and consists of 10 turns. This gigantic loop gives superb results with little fading throughout the shack and some distance beyond. Imagine monitoring the local chat frequency on 2m while watching TV with no other member of the household hearing the calls!

Fig 1. The loop antenna for the hearing aid loop-phone

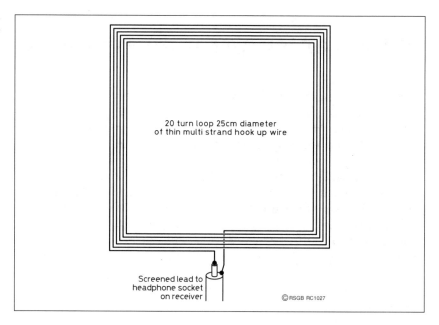

20 turn loop 25cm diameter
of thin multi strand hook up wire

Screened lead to
headphone socket
on receiver

©RSGB RC1027

The most useful loop is the one hung round the neck; it takes the place of the usual headphones. The larger loops can be engineered to suit the shack and house layout. The mind boggles at the thought of a loop wound round the whole house to give complete coverage. Mine consists of 20 turns of hook-up wire wound round itself so that it has a diameter of about 25cm, the two ends of which are connected to the end of a length of screened cable and a plug to fit my headphone socket.

A frame antenna for HF

A receive-only antenna does not have to be efficient to be effective. The most important attribute for receiving is the signal-to-noise ratio. Because the noise source is usually localised, a directional antenna can often be used to remove noise, leaving the required signal in the clear.

THE PROBLEM

Tuned ferrite rod antennas can be very good at achieving this but I have always found that on HF they are not as effective as they are on the LF bands. The frame antenna is far superior and you do not have to find suitable ferrite!

THE SOLUTION

To achieve a wide tuning range, the usual solution is to have a large coil of wire and select sections of it using a switch. A variable capacitor is then used to resonate the coil on the band in use.

My solution was to make the loop resonate at several frequencies at the same time. The circuit diagram is shown in Fig 1. C1, C2 and C3 are the three gangs of a 500pF tuning capacitor, as found in older broadcast receivers and at rallies. L1, L2 and L3 are sections of a single-frame coil as shown in Fig 2 and tapped at two points. The circuit's behaviour is very complex, with at least five resonant frequencies for any given setting of the capacitor, but the three main ones cover the amateur bands.

The tuned circuit formed by L3, C2 and C3 is designed to cover the range 11 to 30MHz for a full sweep of the tuning capacitor. L2 with C1 and C2 resonate at about 7 to 15MHz, and L1 with C1, C2 and C3 cover 1.5 to 4MHz. It must be expected that even a slight variation in design will vary these bands, but the differences between my Mk1 and Mk2 efforts were slight, the main difference being that Mk1 worked up to 35MHz, whereas the final one just reaches 30MHz.

Coupling the antenna to the receiver is accomplished by a single-turn loop round the outside of the frame antenna.

CONSTRUCTION

This is up to the constructor to a large degree, but I used 15mm square timber, half-slotted to make the cross. I then used six 10mm pins hammered in at the ends of three of the arms, leaving 3mm exposed. The fourth and lowest arm needs seven pins. These pins were used to support the tinned copper wire, which was then wound round

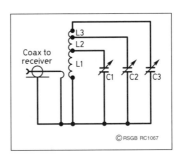

Fig 1. HF frame antenna, circuit diagram

3

Fig 2. General constructional arrangement of the HF frame antenna

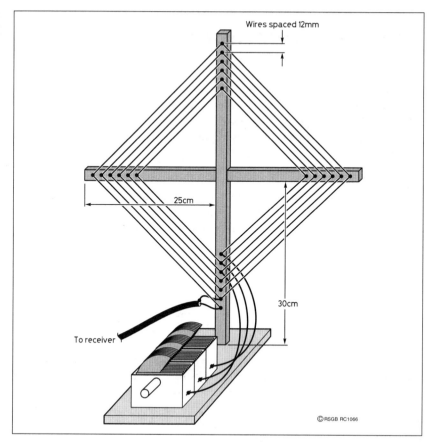

Wires spaced 12mm

25cm

30cm

To receiver

©RSGB RC1066

the pins and soldered to hold it in place, as Fig 2 illustrates. The three-gang capacitor was fixed to the base plate and short lengths of wire used to connect to the frame antenna.

An AM transmitter for 160m

Have you ever thought it would be nice to have a band that would be great for local chatting, good for mobile operating without lots of flutter and loss of signal when going round the far side of a hill, while using cheap, homebrew gear? Area coverage during the day is greater than on 2m and 70cm, because 'dead spots' are penetrated to a degree, and at night the range increases dramatically, with coverage of the whole UK possible. Sounds ideal, doesn't it? Then read on . . .

BACKGROUND

We have such a band and it has nearly been forgotten because early Japanese commercial radio equipment did not have it and it went out of fashion. When I started in amateur radio, 95% of newly licensed amateurs cut their teeth on this band. They learned good operating practices and realised that power had to be used efficiently because the maximum input power to the RF power amplifier was only 10W.

The band is known affectionately as *top band* and covers 1.81 to 2.0MHz. This band is very popular with radio amateurs who construct and operate amplitude modulation (AM) equipment.

The advantage of AM is that the transmitter (like the one described in this article) is relatively simple. A further advantage is that a BFO (beat frequency oscillator) is not required at the receiver. The signal can be received on a broadcast receiver that covers the lower short-wave band (rather rare) or on a medium-wave broadcast receiver modified to cover top band.

SSB receivers (or direct-conversion receivers) can be used to tune into an AM signal by adjusting the tuning so that the carrier is zero beat and inaudible. This article describes how to build the PA and modulator of a 160m transmitter.

CONSTRUCTION

The transmitter is based on a power amplifier (PA) using a field-effect transistor (FET), driven by a variable-frequency oscillator (VFO). The VFO circuit, which determines the transmit frequency, is shown in Fig 1. No constructional details are given for this. You might want to use the modulator and PA with a VFO you already have. However, the circuit is there if you want to make use of it. L1 and VC1 need to be chosen to cover the 160m band.

For AM, we require an audio amplifier to amplify the microphone signal to a level sufficient to modulate the PA. For this, I have adopted an idea by Doug Gibson, G4RGN, see Fig 2, used in a design called 'The Poppet', which he published in the G QRP club magazine *SPRAT*. In this design, an audio amplifier integrated circuit (IC) is used to amplify the audio signal so that it will fully modulate the PA. The voltage present on the output pin of an

Fig 1. Variable frequency oscillator circuit diagram

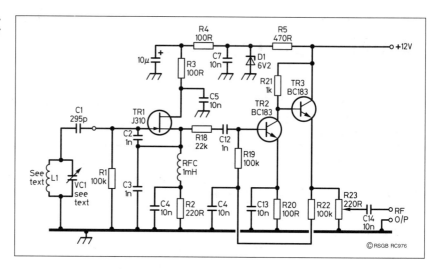

Fig 2. Power amplifier and modulator circuit diagram

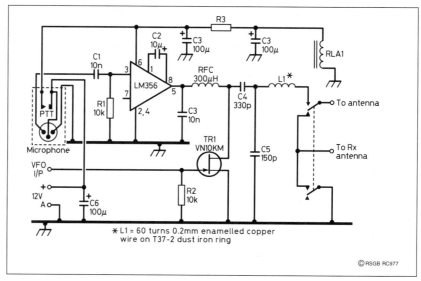

audio amplifier is around half the supply voltage and is able to supply considerable DC current. When a microphone signal is fed to the amplifier the output voltage will swing from near zero to almost supply voltage – this is what is required to amplitude modulate the power amplifier. Using this method means that the PA had to operate on a 6V supply but it does greatly enhance the ease of construction. My transmitter has an output power of 500mW and is modulated up to nearly 100%.

The VFO is mounted in a solid metal die-cast box for mechanical rigidity and good screening. The driver, PA, and output matching and filtering are built 'dead-bug style' on a piece of PCB material; see Fig 3. The antenna changeover between transmit and receive uses a miniature relay. For simplicity, transmit/receive switching uses the switch in the microphone to switch the supply to the transmitter. A 'bypass' switch to the VFO circuitry

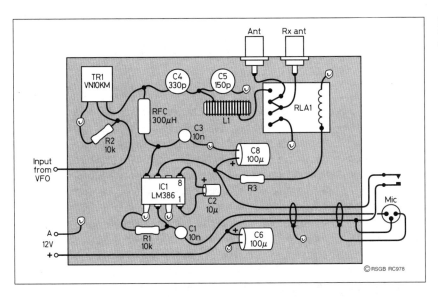

Fig 3. Power ampli-
fier and modulator
component layout

Table 1. Components list for the power amplifier and modulator

Resistors

R1, 2	10k, ¼W
R3	120R, used if 5V or 6V relay fitted

Capacitors

C1, 3, 7	10n
C2	10μ, 25V electrolytic
C6, 8	100μ, 25V electrolytic
C4	330p ceramic
C5	150p ceramic

Inductors

L1	60 turns of 0.2 enamelled copper wire on a T37-2 dust iron ring
RFC	300μH

Semiconductors

IC1	LM386
TR1	VK10KM

Additional item

Relay	2-pole, 2-way 12V

is required to enable the VFO to be turned on during receive so that it can be 'netted' to the received signal.

The modulator can be used with either a two- or three-terminal-type electret microphone. If you use a two-terminal microphone, a 10kΩ resistor should be connected in the microphone casing between the live microphone terminal and the +12V supply to the switch.

An automatic tape switch

If you are a keen SWL or scanner enthusiast, it is probable that at some stage the need has arisen to monitor traffic on a particular frequency for long periods of time. Doing so can be rather tedious, especially if activity on the chosen frequency is sparse. It is therefore desirable to have a means of automatically recording to tape any transmissions for playback later on.

The circuit described here is designed for this purpose though it has a wide range of applications. It is described here as an external unit, but it could be installed inside a tape recorder if desired.

It has been designed to connect to the 'remote' socket found on many tape recorders. Obviously, the receiver should have a squelch control to prevent audio reaching the unit when no transmissions are being picked up. When the transmission has been completed the circuit produces a delay before the tape recorder is turned off again.

HOW IT WORKS

This project has been designed to run off a 9V or 13.8V power supply. This is so that it may be used with a standard low-current power supply. If you do not wish to use a mains power supply, a rechargeable battery pack using AA or larger batteries is recommended.

The circuit is built on a piece of Veroboard cut to 19 holes × 14 tracks. The layout is shown in Fig 1.

The circuit diagram for the unit is shown in Fig 2. The audio from the receiver is amplified by TR1. This is an npn transistor; although I have specified a BC108 in the components list for the prototype, almost any npn transistor will work. The amplified audio is fed to the inverting input of the comparator formed by the 741 IC, an operational amplifier. The voltage at the non-inverting input to the comparator is set by the potential divider RV1.

When the amplified waveform from TR1 is positive-going, the voltage at the output of the comparator falls from the supply voltage to around 2V. As a result, a square wave is produced. This wave is smoothed by C3, which also provides the delay which enables the relay to remain on for about two seconds when the audio signal is no longer present. TR2 cuts off TR3 which in turn energises the relay. This arrangement provides a smooth DC signal to the relay, eliminating jitter.

The relay and D1 are not mounted on the board. This is to allow the use of any 9V or 12V relay according to the desired supply voltage.

CONSTRUCTION

Before mounting the components, the breaks in the tracks should be made as shown in Fig 1. Before mounting IC1 on the board, pins 1 and 5 must be removed. These pins are not required by the circuit and if not removed could prevent the circuit from functioning properly. [This is not recommended practice but, if you intend using the author's strip-board layout, it is necessary – Ed] Take care to insert the three transistors correctly. Remember that the capacitors are electrolytic and must also be inserted with the correct polarity.

When all of the components have been inserted, the next step is to connect the relay. The two wires from tracks A and I as shown are to be connected across the relay coil pins. D1 must also be connected across these two pins as shown with the cathode (striped end) connecting to the wire from track A. Failure to do otherwise could damage TR3.

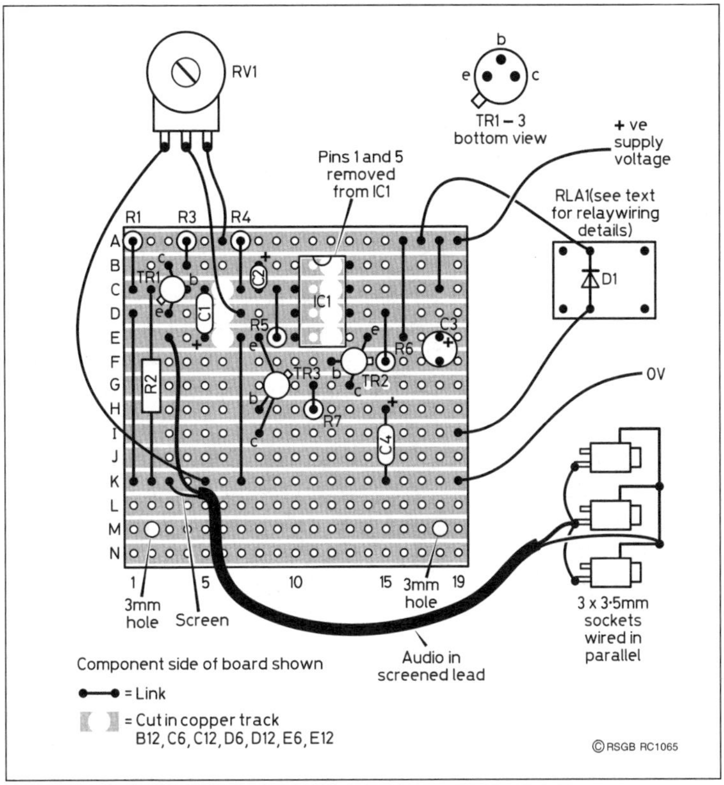

The plug for the 're-mote' socket on the tape recorder is connected to the common and normally open contact pins on the relay, ie the two pins which make contact when a current flows through the coil. As an alternative, these two pins could be wired to a socket on the front panel of the unit. A patch lead can then be used to connect the two sockets as described later; this is my preferred method.

I also mounted a further three sockets on the front panel which are wired in parallel. These are connected to the audio input wires to the board as shown in Fig 1. This allows connection of the radio audio output through

Fig 1. Component layout of the tape recorder VOX unit

Fig 2. Circuit diagram of the tape recorder VOX unit

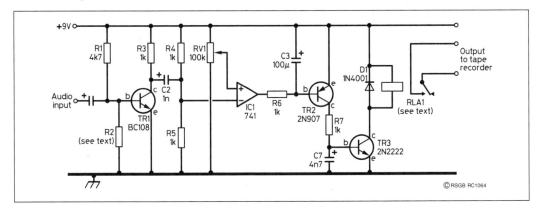

Table 1. Components list

Resistors
R1   4k7
R2   270R for 13.8V supply
      or 390R for 9V supply
R3–7 1k
RV1   100k

Capacitors
C1, 2 1μ
C3   100μ
C4   4μ7

Semiconductors
IC1   741 operational
      amplifier
TR1   BC108
TR2   2N2907
TR3   2N2222
D1   1N4001

Additional items
12V relay for 13.8V power
supply or 9V relay for 9V
supply
Veroboard, 19 holes × 14
tracks
Veropins if required
4 × 3.5mm mono jack plugs
2 × 5mm mono jack plugs
3 × 3.5mm mono jack sockets
2.5 mm jack socket
Suitable case
Wire, solder etc

the unit to the tape recorder microphone socket. The third socket allows headphones to be connected to monitor the audio from the radio if required.

TESTING THE BOARD

Before you begin to test the board for the first time, make sure that all of the components have been inserted correctly and the tracks on the underside of the board are not bridged with solder.

Connect the power supply wires to the power supply. You should find that when the potentiometer RV1 is rotated fully one way the relay will stay on, and if rotated the other way round it will turn off after a short delay.

It is necessary to set RV1 at the threshold, ie at the point where the relay just turns off. To do this you should set RV1 at any point where the relay turns off after the delay. Then very slowly rotate RV1 in the opposite direction until the relay just turns on again.

When it does so, rotate RV1 back again slightly. The relay will turn off after the delay and the unit is now set at the threshold. Connecting an audio source from a radio or cassette player to the audio input lead will turn on the relay. You will need to experiment with different receiver volume control settings and possibly different threshold settings.

When your board is working as described, it should be mounted in a suitable case. Two holes may be drilled into the board as shown in Fig 2 for mounting. I used double-sided, self-adhesive fixing pads to mount the relay. The four sockets are mounted on the front panel. You could put a power supply socket on the rear for the power supply connection if you wish.

On the prototype, a 2.5mm socket was mounted for the connection of the relay to a similar 2.5mm socket on the tape recorder. A patch lead consisting of a length of suitable coaxial cable terminated with two 2.5mm plugs was assembled to link these two sockets.

The remaining three sockets are 3.5mm. These are labeled 'Audio In', 'Audio Out' and 'Monitor' (for headphones). As these are wired in parallel, the order in which you label them is not important. Two more patch leads terminated with 3.5mm plugs were made, one to connect the unit to the receiver's earphone socket and one from the unit to the microphone socket on the recorder.

When setting the squelch control on the receiver, it should be rotated far enough to eliminate any frequent pops or very weak stations which are of no interest, or you will find the tape fills up with 'junk'.

Please note: The author and the publishers of this article wish to point out that the ownership of the unit described above does not imply any licence or rights to record any radio transmissions not intended for reception by unauthorised persons.

A simple electronic keyer

Fifty years ago, a Danish amateur, OZ7BO, designed a clever and simple electronic keyer which became known as the *El-Bug*. This device used a double triode valve and a few small components. Some 30 years later the design was updated by G3JIS using two transistors in place of the valve. Both designs used two relays, one of which is used to key the valve transmitter, which had a high positive voltage on the keying line.

These days, most transceivers have keying voltages of less than 12V positive to ground and the keying currents are low. It would be best to check your keying line prior to starting this keyer and, if it is a negative keying line or uses high current, a second relay could be fitted to the keyer output to key the transmitter.

The low power requirement of modern sets enables us to simplify the G3JIS version to 12 components, plus the bonus of a cheap and easily obtainable relay. If all the parts are purchased new the total cost will be about £3.

The El-Bug will not send automatic CQs or make the coffee but, handled properly, it will send perfect Morse at any reasonable speed after a little practice. It can be used with either a single- or double-lever paddle, but will not be *iambic* (ie it does not send alternating dots and dashes if both paddles are held together) and can be built on a 5cm × 6cm board without overcrowding.

CONSTRUCTION

I use a surface-mount assembly method, which is a cross between the saw-cut technique advocated by G3VTS ('TT', *RadCom* April 1995) and 'dead-bug' construction favoured by G3ROO. All parts are mounted on the copper side of the board and the soldering pads are made large which simplifies construction. This allows for modification to be made with ease. To make a board like this is very easy indeed as no advance planning is necessary providing that more pads are made than you think will be needed.

Start by sticking PVC tape across the board in parallel lines, with 2 or 3mm gaps between them. Next, remove the unwanted tape with a sharp knife and the board is ready for etching in the normal way. No holes are required, as parts like relays and ICs can be stuck to the board 'dead-bug' fashion. The result will be as neat or ugly as you wish but I have built an SSB transceiver using eight boards constructed by this method and they have proved to be invaluable when modifications have been necessary.

At first sight it seems hard to comprehend how such a simple circuit, shown in Fig 1, can generate dots and dashes of correct length and spacing without a binary digit in sight. To understand the operation, first note that the relay contacts used are in the normally closed position. When the paddle makes contact with the dash side the supply voltage is applied to C1

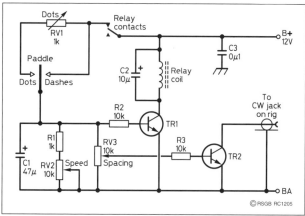

Fig 1. Circuit diagram of the modified OZ7BO autokeyer

which charges, applying a positive voltage to both transistor bases. TR1 saturates, energising the relay and opening the contacts. At the same time TR2 saturates and keys the transmitter. With the supply to C1 removed by the relay contacts it discharges via the complex network connected across it. As the base voltage of TR2 drops below 0.6V, TR2 turns off and the transmitter is unkeyed. At this moment, TR1 is still on due to its base voltage being slightly higher than that of TR2. As C1 continues its discharge, TR1 turns off and the relay closes again ready for the next operation. If the paddle is kept pressed a stream of dashes will be sent until the paddle is released. The difference is spacing is set by the time difference in the off states of TR1 and TR2. Due to the setting of RV3, TR2 is always switched off slightly earlier than TR1. A similar cycle takes place on the dot side, except that the quantity of charge in C1 is restricted by RV1, thus shortening the dashes to dots.

Correct operation of the El-Bug is very dependent on the characteristics of the relay. In fact it is the mechanical sluggishness of the relay that enables the circuit to operate at all. A 12V relay with a coil resistance of 400Ω is ideal. I use a miniature type YX94C from Maplin. This one is a little fast in operation, so I have retarded it with C2 connected across the coil. The value of C2 with the Maplin relay is not too critical, but a junk box relay may need some experimentation to get correct operation and a slower relay may not need any capacitance. The values of C1 and the three variable resistors are critical; do not leave out R1 or you may short out the power supply!

For initial testing, set the speed pot RV2 at half-track, which should correspond to about 12WPM. With an ohmmeter, set RV3 for 3kΩ between slider and ground. Connect a 12V supply, hold the paddle in the dash position and you will hear the relay clicking away quietly. Connect your rig into a dummy load and connect the key to the keying line. On pushing the dash paddle, dashes should be heard on the sidetone; adjust RV3 for optimum mark/space ratio. Now key the dot paddle and adjust RV1 for correct dot mark/space ratio. With the components specified, the keyer should work between 8 and 25WPM.

To prevent RF problems, the keyer should be built in a metal box and coaxial cable used to connect to the key jack of the transmitter. I had problems on 15m but this was cured by fitting two 10nF ceramic capacitors, one from the collector of TR2 to ground and the other between its base

Table 1. Component list

Resistors

R1	1k
R2, 3	10k
RV1	1k preset
RV2	10k linear potentiometer
RV3	10k preset

Capacitors

C1	47μ, 25V
C2	10μ, 25V
C3	0μ1, 60V

Semiconductors

TR1, 2 BC108, BC109 or BC171 or near equivalent

Additional item

Relay (Maplin YX94C)

and ground. With these fitted, full-power operation was available on all bands.

OPERATING

A newcomer to auto-keyers may well find first efforts depressing, as the keyer seems to have a mind of its own. I remember 10 years ago borrowing a keyer for a few days and, although the marvel of it was obvious, I was unable to master it in five minutes and gave up. Now the beauty of making your own for a few pounds is the great psychological advantage you gain. Because you made it yourself and because it actually works, you feel obliged to persevere. You may well give up and put it away for a while but, sooner or later, you will go back for another go.

After three and a half hours' practice, I was able to do practice-QSOs at 12WPM and soon ventured on the air. I made mistakes, but slowly got better and the El-Bug has revitalised my interest in Morse and makes sending a pleasure. Paradoxically, sending good Morse has improved my receiving as well.

Field-strength measurement

Never expect the same antenna to work the same in two locations. Ground conductivity and reflections from objects, even at considerable distances, can upset our expectations. The effects are more pronounced on VHF, and experimentation with a field-strength indicator (not 'meter', as no accuracy can be claimed) can be extremely enlightening.

USE OF THE INDICATOR

You can get some idea of the field-strength pattern of your antenna by taking a reading on your completed field-strength indicator, then walking around and noting how the reading changes.

CONSTRUCTION

This indicator was originally designed for a specific use: testing glider radio systems. Because the indicator has subsequently proved so useful for all amateur radio bands from 160m to 70cm, it is worth passing on.

Several designs were tried to begin with, the most interesting being a dipole with a 68Ω resistor connected between the two halves and an RF detector connected across the resistor. With the dipole securely mounted on a tripod, the dipole length could be adjusted for maximum reading, enabling the wavelength of the signal to be measured with surprising accuracy.

Unfortunately the sensitivity of the unit was not sufficient for the purpose so the resistor and detector were replaced by a full-wave bridge as shown in Fig 1.

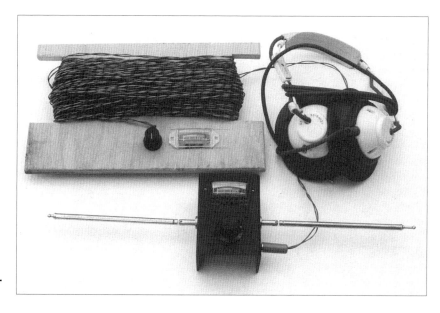

The completed indicator ready for use

A wide-band amplifier could have been added to the original circuit to gain the necessary sensitivity, but experience has taught me that it is far better to use a 'passive' unit than one requiring a power source, as the battery has the annoying habit of being flat when needed most urgently.

In the circuit diagram there is a jack socket labelled 'phones'. This has two functions. On the aircraft band, amplitude modulation is used, and the modulation can be checked by plugging a pair of high-impedance headphones into this socket. Unfortunately, FM is the most popular mode used on the amateur VHF bands and this is not rendered audible using a diode detector.

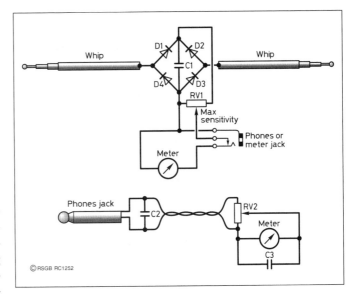

Fig 1. Circuit diagram of the field strength indicator

The second function enables an extension meter to be used. This consists of a second meter and sensitivity control which, when used, requires the sensitivity control in the head unit to be set to maximum. The extension cable can either be two lengths of hook-up wire twisted together or a length of coax. If you have a length of coax in your junk box which has been removed from service as it has become lossy at RF, it would suit this job admirably because a DC measurements are unaffected by the problems that cause the cable to be lossy at VHF.

A simple way to twist long lengths of hook-up wire is to lay out two lengths of wire on the ground and make the far ends fast to a solid anchor such as a tree or fence post. The two free ends are then clamped into the chuck of a hand drill and, with the wires under light tension, the drill is used to twist them together.

The SWR bridge is very useful when tuning matching networks at the base of antennas but can be misleading at times. Spurious resonance within the matching network can, on occasions, indicate a good VSWR when there is a poor match to the antenna. Using a field-strength indicator on a tripod some distance from the antenna and an extension meter alongside the operator enables quick and easy tuning-up of systems.

There is little to be said about the construction of the indicator, with the exception that all wires should be kept as short as possible. The meter recommended is a 50µA unit. However, a 200µA meter from the junk box was used; this reduced the sensitivity but was very cheap!

Table 1. Components list	
RV1, 2	10k linear potentiometer
C1–3	10n ceramic
D1–4	OA2 germanium diode or similar
Meter, 50µA FSD	
Antenna halves, 560mm telescopic whips	
Jack socket with switch and plug to match	

A bi-directional wattmeter

Three members of a radio club are building 80m CW transceivers for holiday use. A power/SWR meter was needed that could operate at low power and did not need to be adjusted every time the frequency or power was changed. This is the design that was chosen.

HOW IT WORKS

When the transmitter is connected to P1 (Fig 1) and the antenna to P2, 99% of the transmitter power arrives at P2 to go to the antenna. The other 1% is sampled by the transformers and fed to R1 and R2. The RF voltage across R1/R2 is rectified by D1 and fed to meter M1. The phasing of the windings on transformers T1 and T2 cancel any voltage on R3 and R4.

If the antenna does not present a perfect 50Ω impedance, some power will be reflected. This reflected power is 180° out of phase with the transmitted power. About 1% of this reflected power is sampled by the transformers and fed to R3 and R4. The RF voltage across R3 and R4 is rectified by D2 and fed to meter M2. The phasing of the windings on transformers T1 and T2 cancel any voltage on R1 and R2.

CONSTRUCTION

The power meter RF components must be constructed in a metal box (or a box made from PCB material), as shown in Fig 2. The transformers are wound on ferrite beads (Fair-Rite 26-43006302, Bonex 6302 PA balun bead), which are fairly large and are easy to handle.

Each transformer has 14 turns of 26SWG enamelled copper wire wound tightly on the core with the turns equally spaced. Ensure that both transformers are wound with the same number of turns. The second winding consists of a short length of thin coax threaded through the centre, one end of its braid connected to ground, and the other neatly trimmed off and left disconnected.

The braid screens the two windings as far as the electrostatic field is concerned but will not hinder the electromagnetic field. (This is known as a *Faraday cage* or *screen*).

If 'junk box' meters are used, it will be necessary to calibrate the unit using a known power meter. This is not difficult to do,

Fig 1. Circuit diagram of wattmeter

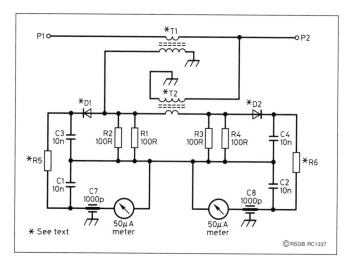

16

provided your transmitter can be reduced progressively in output to about 1W. This is the procedure to follow.

CALIBRATION

Connect the transmitter output to your power meter, the output of which must then be connected to the reference power meter which in turn must be connected to a 50Ω dummy load. A 100kΩ variable resistor is connected in place of R5 and the transmitter adjusted to give an output of 10W, as indicated on the reference meter. The 100kΩ pot is then adjusted for full-scale deflection (FSD), its value is measured and R5 and R6 replaced by resistors of the correct values. By progressively reducing the output power a graph of calibration can be made for the meter and from that a new meter scale can be laid out.

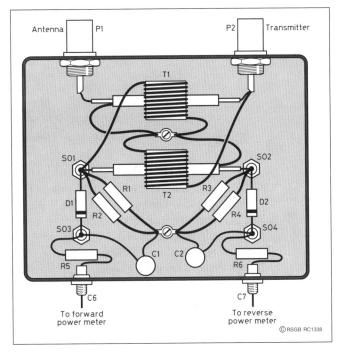

Fig 2. Component layout (meters not shown)

Making scales is easier than most people realise, provided a photocopier with a zoom facility is available. Remove the old scale and photocopy it onto paper at four times its original size. On the photocopy, the new scale is then carefully marked over the old scale using Letraset figures and lines, and all unwanted markings are covered with Tipp-Ex. Now, rather than reducing the size by four on the photocopier, reduce it by two and then touch out any remaining imperfections using Tipp-Ex, then reduce by two again to regain the original size. Using either spray-on adhesive or thin double-sided tape the new scale is stuck over the old scale and re-mounted in the meter casing.

This power meter is designed for 10W, but the 43 mix beads can be used up to 100W without problems. Calibration is carried out the same way but it is necessary to increase the power rating of R1, 2, 3 and 4. Two 100Ω ⅛W resistors were used to give 50Ω at ¼W for 10W. For the 100W test, four 100Ω resistors were connected in series/parallel to give 50Ω at ½W. On SSB they do get warm, so if SSTV or very-high-duty-cycle modes are to be used, it may be advisable to increase the power rating further.

Table 1. Components list
Resistors
R1–4   100R carbon or metal film
Capacitors
C1–4   10n disc ceramic
C7, 8   1000p feedthrough
Additional items
T1, 2   14 turns of 26SWG copper wire on a Fair-Rite 26-43006302 ferrite ring
For other components see text
The address of Bonex is: 12 Elder Way, Langley Business Park, Slough, Bucks SL3 6EP. Tel: 01753 549502; fax: 01753 543812.

Two-element, 6m quad antenna

6m – the 'magic band' – where you can work across town or try your hand at some real DX during the summer months, particularly while we are still near the sunspot maximum. This simple antenna is robust – more so than a three-element beam in an exposed, windy situation, and it performs well.

CONSTRUCTION

This is simple and uses a chunk of 5mm aluminium plate, nylon cable ties and glassfibre rods. Fig 1 details the construction.

Having drilled the metal plate and loosely fixed the rods in place with cable ties, holes are drilled about 6mm from the ends of the rods and 90° from the plane of the loop. By doing this the wire will not slip when under tension, keeping the symmetry of the loop. The wire lengths are carefully measured, adding 30mm for soldering the ends. The loops are made up from normal multi-strand hook-up wire, although thicker wire could be used without materially affecting the performance. The driven loop and reflector have wire lengths of 543cm and 594cm respectively.

A gamma match is made up on a piece of Perspex using the earth wire out of 2.5mm mains cable for the loop. The wire is folded into a 'hairpin' 31cm long and with 2cm spacing (Fig 2). The thickness of the wire and its small size enables it to be supported only by a piece of Perspex at the capacitor and feed point. About 20pF is required for the capacitor, so a 35 or 50pF variable is used to allow for variations.

The reflector loop is fed through the holes in the ends of the four longer rods and the two ends are soldered together. Next, the driven wire is attached to one side of the gamma match, the free ends passed through the holes of the shorter rods and the loop completed by soldering. In the final quad the gamma match capacitor is in a small plastic box to shelter it from the rain.

Finally the rods are slid in or out of the cable ties so the 700mm loop spacing is correct and then the cable ties are pulled tight, locking the structure in a stable,

Fig 1. Glassfibre spreaders are fixed to an aluminium plate with cable ties

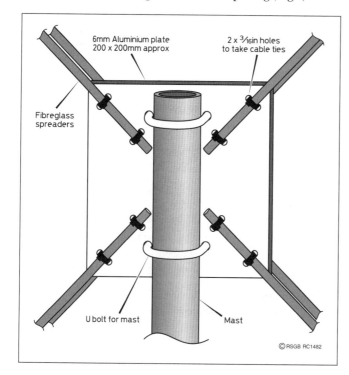

6mm Aluminium plate 200 x 200mm approx

2 x ³⁄₁₆in holes to take cable ties

Fibreglass spreaders

U bolt for mast

Mast

© RSGB RC1482

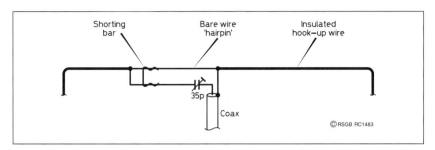

Fig 2. Detail of the gamma matching system

Table 1. Components list
Glassfibre rods, 8 or 10mm diameter, 4 × 108cm long and 4 × 111cm long Aluminium plate, 20 × 30 × 0.5cm 2 × 50mm U-bolts. Car exhaust clamps are ideal 12 × 5mm nylon cable ties. Make sure you have spares Variable capacitor, 50pF Copper wire, 1m solid, 2.5mm Multi-strand hook-up wire, 12m

if not rigid, configuration. It can be made more rigid by tying the ends of the driven loop to the corresponding reflector loop using fishing line.

SETTING UP

Various feed systems were used, the best using 72Ω twin feeder, but the complexity of the balanced ATU in the shack left a lot to be desired, so the second best feed using a gamma match is described.

The gamma match consists of a coupling wire and capacitor. The tapping point and capacitor are adjusted to give a perfect match to the feeder, measured on an SWR meter. A match can be maintained provided the feeder is kept perpendicular to the horizontal part of the driven element and equidistant to the vertical sections. During testing and setting up, you will see the VSWR rise rapidly if the feeder is moved from one side to the other, demonstrating the importance of keeping antenna systems symmetrical! The gamma match may be adjusted by standing on a step ladder with the base of the quad about 8ft above ground. A shorting bar of wire is moved 5mm at a time along the hairpin, starting from the closed end, and the capacitor re-trimmed for best VSWR. This is repeated until the point of best VSWR is obtained. When mounted on the side of the house, the resonant frequency rose by 100kHz and the VSWR rose to 1.2:1 due to the change in location.

The quad antenna is easy to construct if the right materials are at hand and it is easy to maintain. The glassfibre rods used were spares, but they are available from garden centres for making cloches. Failing this, bamboo canes could be used, but these must be selected for thickness and flexibility so that the final shape is satisfactory.

A switched attenuator

How many times has a signal been too strong for the experiment you wish to carry out? It could be from an oscillator on the bench or from signals from an antenna overpowering a mixer. This attenuator will solve those problems.

WHERE CAN I USE IT?

Applications other than those given above are if you are interested in DF and need to attenuate the signal when getting close to the transmitter, or if you have a problem with a TV signal being too strong and causing ghosting on other signals (cross-modulation).

ATTENUATORS

These problems can be averted by using a switched attenuator (pad). The times that we need attenuators occur far more often than first realised.

When designing the attenuators, account must be taken of the distinct possibility of poor screening. There is hardly any point in designing a 20dB attenuator when the leakage around the circuit is approaching this value. It is also important to decide on the accuracy required. If it is intended to do very accurate measurements the construction has to be impeccable, but for comparisons between signals it would be possible to accept attenuation values to a smaller degree of accuracy.

The completed attenuator

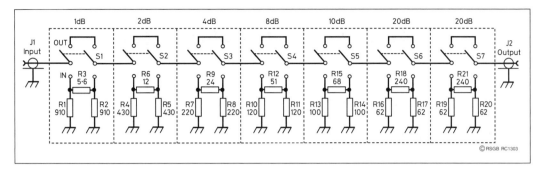

The most useful attenuator is a switched unit where a range from zero to over 60dB in 1dB steps can be covered. This is not as difficult as it first seems because, by summing different attenuators, we can obtain the value we need. It takes only seven switches to cover 65dB. The seven values of attenuation are 1dB, 2dB, 4dB, 8dB, 10dB, and two at 20dB; these can be switched in or out at will. As an example, if 47dB were needed, switch on the two 20dB pads plus the 4, 2 and 1dB pads. [A 'pad' is the name given to a group of components with a known attenuation. – Ed]

Fig 1. The attenuator consists of seven pi-network sections, so-called because each pad (eg R1, R3 and R2) resemble the Greek letter pi (π). Input and output impedances are 50Ω

CONSTRUCTION

The prototype attenuator is shown in the photo and in Fig 1. It was constructed 20 years ago and it is still in regular use. It is housed in a box made from epoxy PCB material. The top and sides are cut to size and soldered into a box.

It is easier to cut the switch holes prior to making the box. After the box has been constructed, screens made from thin brass shim should be cut and soldered between the switch holes.

Next, the switches are fitted and the unit wired up. When this is done, the unit is checked and a back cover, securely earthed to the box, is fitted.

COMPONENTS

The switches must have low capacitance between the contacts and simple slide switches are the best selection. The Maplin DPDT miniature switch (FH36P) would be suitable and Maplin also supplies 1% resistors. Connectors to the unit must be co-axial but can be left to personal preference.

The resistor values shown in Fig 1 determine the attenuator's accuracy at around 5%. This is done for practical reasons. For example, if we wanted to make the attenuation value of the 4dB cell *exactly* 4dB, the resistor values would have to be 220.97Ω and 23.85Ω. You will see from Fig 1 that the values used are 220Ω and 24Ω, giving an attenuation value of 4.02dB.

Table 1. Components list	
Resistors	
R1, 2	910R
R3	5R6
R4, 5	430R
R6	12R
R7, 8	220R
R9	24R
R10, 11	120R
R12	51R
R13, 14	100R
R15	68R
R16, 17, 19, 20	62R
R18, 21	240R
All resistors ½W or 1W carbon or metal film	
Other items	
S1–7	double pole miniature switch
J1, 2	SO239 sockets (or similar to suit your equipment)

An absorption wavemeter

In the absence of another calibrated HF receiver, the absorption wavemeter is a requirement of your amateur radio licence. You need some way of checking the frequency of your transmissions, principally to reassure yourself that you are not radiating outside the specified frequency range, either at your fundamental frequency or any of its harmonics.

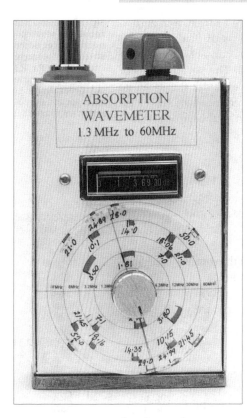

Front view of the absorption wavemeter. The amateur bands are marked in red on the frequency dial

WHAT IT WILL DO

An absorption meter is a simple instrument comprising a variable tuned circuit, detector and a meter. With the capacitor dial calibrated in frequency, this simple instrument can be used for the following measurements:

- indicating the approximate frequency and relative power output of a transmitter;
- indicating the approximate frequency and relative level of significant harmonic outputs from a transmitter;
- checking to see if an oscillator is oscillating;
- measuring relative antenna field strengths, so that the relative performance of antennas or ATUs can be checked;
- plotting rough polar diagrams of beam antennas.

The absorption meter described here covers the frequency range 1.8 to 60MHz in four ranges.

CONSTRUCTION

The circuit is shown in Fig 1. The tuned circuit is fabricated from home-made coils and a polyvaricon capacitor. As you can see from the photos, the absorption wavemeter was built in a metal box, although a plastic one would be equally suitable. A piece of PCB material, cut to size to fit in the box, acts as a 'sub-chassis' on which to solder the components as shown in the photo. The chassis provides a method of mounting the polyvaricon capacitor, which is designed for PCB mounting – its connections are to the four pins on the shaft end of the casing. The tags at the rear end are connected to the internal trimmer capacitors, which are not used in this unit. Only one of the two 310pF variable capacitors is used in this design, the other 310pF and the two 22pF capacitors are left disconnected. The capacitor 'frame' connections are to the two large solder tags on the shaft end of the case.

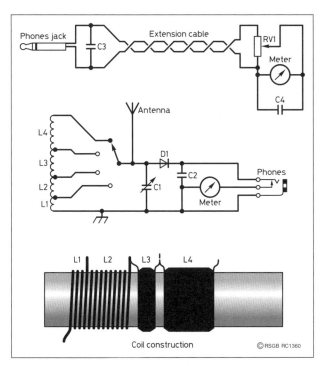

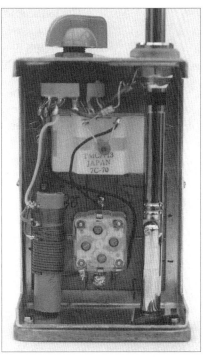

Coil construction © RSGB RC1360

Table 1. Components list

Resistors
RV1   10k linear potentiometer

Capacitors
C1   310p polyvaricon
C2–4   10n disc ceramic

Inductors
See Fig 1

Semiconductor
D1   OA81 or similar *germanium* diode

Additional items
Telescopic antenna
50µA FSD meter
Phone jack to suit available headphones
Rotary switch, single-pole, 4-way

The PCB is drilled for clearance of the shaft and to mount the polyvaricon. The positions of the capacitor pins are carefully marked with a felt-tip pen and then the copper is cut through with a model drill or broken hacksaw blade, making insulated islands to which the pins are soldered. The rest of the construction is straightforward and should present little difficulty.

The coil is wound on an old felt-tip pen case, holes drilled to enable tinned copper wire tags to be threaded through to anchor the ends of the coils in place. The coils are all wound in the same 'sense', ie in the same direction. The inductances will then add together, enabling the coils to be used in series. For Range 1, L1 only is used; for Range 2, L1 and L2 are used, and so on. The coils use different diameters of wire but in practice alteration of wire diameters of up to 20% will make little difference, and the range-overlaps should cope with the error.

As you can see in Fig 1, the design incorporates a headphone socket, which will enable the operator to listen to amplitude modulation if required.

Left: Fig 1. Circuit diagram of the absorption wavemeter. The coils of enamelled copper wire are wound on a 10mm felt tip pen case and the windings are: L1, 3.5 turns of 0.9mm (20SWG); L2, 10 turns of 0.9mm (20SWG); L3, 21 turns of 36SWG pile-wound over 4mm coil former length; and L4, 50 turns of 36SWG pile-wound over 10mm coil former length

Above: Rear view. The PCB sub-chassis can be seen behind the components, fixed to the front panel

The telescopic antenna was obtained from an old transistor radio. If you use a metal case, you will need to make the antenna mount out of a section of insulating material.

Using new components, the cost was less than £20 but could be reduced considerably by using components from the junk box. The polyvaricon capacitor could be obtained from an old transistor radio but look carefully at its value. For example, one of the older polyvaricon units contained two 180pF variables; these would be used in parallel, making 360pF.

The old rustic paddle

Paddles for electronic keyers are usually a mechanical arrangement, often constructed with watchmaker precision. For portable or mobile operation such a paddle is easily damaged. Is there an alternative? Try this one . . .

THE RUSTIC PADDLE

This paddle has no springs or delicate mechanical parts. It comprises pieces of PCB material and flexible insulating material clamped between wooden blocks, and then fixed to a baseboard. The complete paddle is shown in Fig 1.

CONSTRUCTION

The paddle arm is made from two shaped pieces of PCB material stuck together back-to-back; see Fig 2. The dot and dash contacts are made from pieces of PCB stuck to blocks of hardwood ($65 \times 30 \times 15$mm) using double-sided adhesive tape. Having done this, the three pieces are held together in their final position and two pilot holes are drilled to take the screws.

The holes in the three pieces of PCB and one in the wooden block are enlarged and elongated to clear the screws when they are inserted. Two pieces of dense sponge rubber (wetsuit neoprene is ideal!) are stuck as shown in Fig 2 to the contact PCB. This makes construction easier, and the unit can then be assembled with the screws.

The wooden side blocks are fitted to a sub-base block; see Fig 2. Pilot holes are drilled in the sub-base block for the screws to hold the wooden side blocks.

The front holes in the sub-base block are enlarged; this is to allow movement for the paddle spacing adjustment.

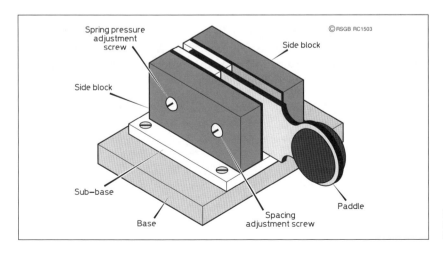

© RSGB RC1503

Spring pressure adjustment screw

Side block

Side block

Sub−base

Base

Spacing adjustment screw

Paddle

Fig 1. Construction of the complete paddle key

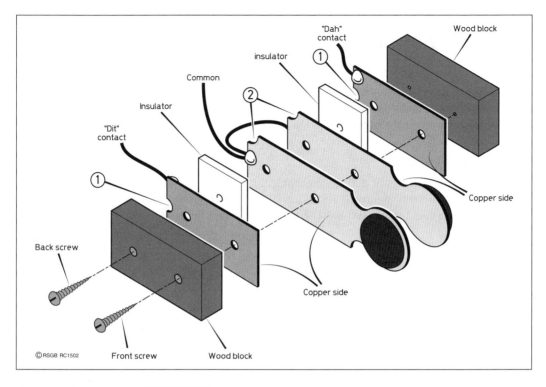

"Dah" contact
Wood block
insulator
Common
Insulator
"Dit" contact
Back screw
Copper side
Copper side
Front screw
Wood block
© RSGB RC1502

Fig 2. Exploded view of the paddle components. Cutouts (1) and (2) are to allow room for the solder blob connections on the facing PCB plates

ADJUSTMENT

Fix the paddles to the sub-base board but do not tighten fully. Connect the three wires to the electronic keyer. Adjust the 'spring pressure' using the rear screw, as shown in Fig 1, and the spacing using the front screw. These adjustments are interactive and may have to be repeated.

Tighten up the screws holding the paddles to the sub-base board and check the 'spring tension' and the spacing again. Fix the sub-base to a main baseboard as shown in Fig 1.

FINALLY

I used my paddle in my mobile station as its construction is robust and cheap, and it is ideal for mobile or portable use. With a little more complexity (ie the two parts of the paddle operating independently) this idea could be converted for use with an iambic keyer.

A T-match ATU

An antenna tuning unit (ATU) is very useful if you are a licensed radio amateur or a short-wave listener. The purpose of an ATU is to adjust the antenna feed impedance so that it is very close to the 50Ω impedance of the receiver or transmitter, a process known as *matching*. When used with a receiver, an ATU can dramatically improve the signal-to-noise ratio of the received signal. On transmit, the antenna must be matched to the transmitter so that the power amplifier operates efficiently.

DESIGN

The basic ATU design is called a *T-match*; you can see the basic shape of the letter 'T' reflected in the layout formed by the components VC1, VC2 and L1 in Fig 1. The circuit will match the coaxial output of the transceiver to an end-fed antenna or to a coaxial cable feed to the antenna. This design also uses a *balun* (*bal*ance to *un*balance) transformer for use with antennas using twin-wire balanced feeder.

This unit will handle up to 5W and operates over the frequency range of 1.8 to 30MHz.

CONSTRUCTION

Inductor L1 is wound on a T-130-2 powdered-iron toroid. The inductor is tapped and fixed to the tags of a 12-way rotary switch. Taps are formed by making a loop about 1cm long in the wire and twisting it tightly. The loops are scraped clean of enamel and tinned with solder ready to be soldered onto the tags of the 12-way rotary switch. If the loops are about 1cm long, it is just possible to bend them to fit the tabs of the switch without having to extend them with short wires.

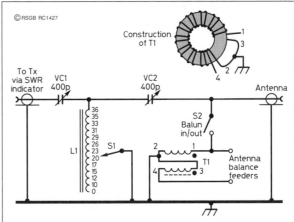

The balun is wound with two wires twisted together; known as *bifilar winding*. These two wires can be twisted together (before winding on to the toroid) by fixing one pair of ends in a vice, the other ends in a small hand-drill. The drill is then slowly rotated so that the two wires are twisted together

Top: The completed ATU

Above: Fig 1. Basic circuit of the ATU

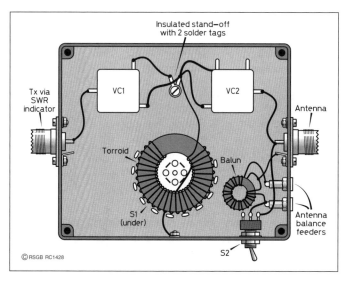

Insulated stand–off
with 2 solder tags

Tx via
SWR
indicator

VC1

VC2

Antenna

Torroid

Balun

S1
(under)

Antenna
balance
feeders

© RSGB RC1428

S2

Fig 2. Internal view of the ATU. A small metal box makes a suitable housing

Table 1. Components list

Capacitors
C1, 2   200 × 200p

Inductors
L1   T-130-2 powdered iron toroid with total 36 turns of 22SWG enamelled copper wire, tapped at 10, 12, 15, 17, 20, 23, 26, 29, 31, 33 and 35 turns from the earth end
T1   12 bifilar turns of 26SWG enamelled copper on FT-50-43 toroid

Additional items
S1   1-pole 12-way rotary switch
S2   SPST toggle switch

RF connectors, SO239 sockets or similar (see text)
Aluminium or die-cast metal box
Three plastic knobs for capacitors and switch
Two 4mm sockets for balanced antenna connection

neatly. Identify the start and finish of each winding with a buzzer and battery or an ohmmeter. The finish of the first winding is joined to the start of the second, as shown in Fig 1.

The two capacitors are twin 200pF polyvaricon capacitors with both gangs connected in parallel to give 400pF max. You have to drill holes in the box for the control shafts of the capacitors and the switch, and the RF sockets. The switch is fixed using a nut on the control shaft and the two capacitors are fixed using adhesive (hot melt glue is preferred). Take care not to let any tags from the capacitors touch the box as both sides of both capacitors are not earthed.

An appropriate RF socket, such as a SO239, BNC or phono socket may be used – the choice is yours and should suit your existing equipment. 2–4mm sockets may be used for the balanced output. See Fig 2 for layout.

It is a good idea to make a graduated dial for each of the three control knobs. An alternative is to use calibrated knobs. You can then calibrate the settings of the three controls so that they can rapidly be reset when you change frequency bands.

OPERATION

The best indication of optimum matching can be achieved using an SWR bridge; the ATU controls are adjusted sequentially and several times for minimum SWR. If used for receive only, the best antenna-to-receiver match can be achieved by adjusting the controls for maximum signal.

An L-match ATU

Many antenna tuning units (ATUs) have been tried, but good results are regularly obtained from an L-match tuned against quarter-wave counterpoises (elevated 'ground wires'). Actually, only two counterpoises, 20m and 5m long, are needed to cover all bands from 80m to 10m.

CONSTRUCTION

The L-match employs just two main components, a coil and a capacitor, as shown in Fig 1. The coil is wound on a plastic 35mm film container. It has a total of 50 turns of 24SWG enamelled copper wire, wound tightly and without spaces between the turns, as shown in Fig 2. It is tapped at each turn up to 10, then at 15, 20, 25, 30, 35, 40 and 45 turns. The use of crocodile clips permits any number of turns to be selected.

The taps on the coil are formed by bending the wire back on itself and twisting a loop. The loops are then scraped and tinned with solder. Some enamelled copper wire is self-fluxing and only requires the application of heat from a hot soldering iron bit loaded with fresh solder for a few seconds before it will tin. Make a good job of tinning the loops to ensure the crocodile clips make good contact. The capacitor is a polyvaricon from an old radio. I used a 2×200pF component with both gangs in parallel but other values will work.

The ATU was made up on a wooden base with scrap PCB for the front and rear panels, as in Fig 3. You will need a socket to go to the transmitter or receiver and one each for the wire antenna and the counterpoise or earth. You can use whatever matches your existing equipment. The prototype used a phono socket for the transmitter and two 4mm sockets for the antenna and earth. The coil is mounted by screwing the lid of the film container down on the base, then snapping the completed coil assembly into it. I glued the capacitor to the front panel. Wire up point-to-point, as shown in Fig 3.

OPERATION

When completed, you can check operation with a receiver or a low-power transmitter (less than 5W). With a transmitter, use an SWR meter to find the coil tap which gives the lowest SWR, then adjust the capacitor to tune to minimum SWR. Incidentally, remember not to touch any exposed

Front view of the simple L-match ATU, showing crocodile-clip connection to the coil. The knob must be of the insulated type, as both sides of the variable capacitor are potentially live to RF

Fig 1. The L-match is constructed from just two components

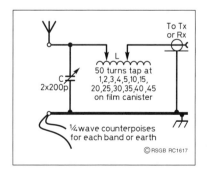

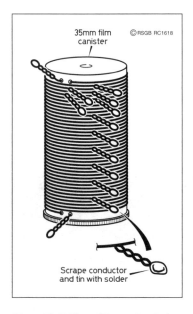

Above: Fig 2. The coil is constructed on a 35mm film container

Above right: Fig 3. Physical layout. The front and rear panels are fixed to the base with wood screws

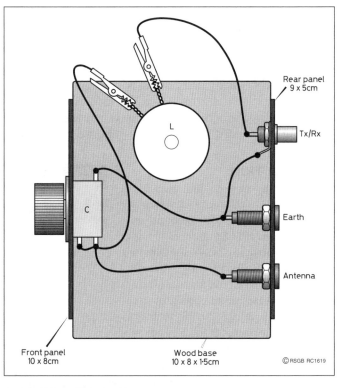

Right: Rear view of the ATU, showing sockets and connections to the capacitor

metal within the ATU while on transmit! An insulated knob and a scale helps you to note the position of the capacitor for future reference. To set it up using a receiver only, find the best position for the tap and the capacitor by listening to a weak signal and adjusting for the loudest signal (with the help of an S meter, if you have one).

As a guide, I used 20 turns on 160m, 10 turns on 40m, 4 turns on 20m, and 2 turns on 10m and 15m.

A signal injector

One of the most satisfying things in amateur radio is being able to locate and repair faults on equipment. One of the most useful tools for fault detection is a signal generator but, as these are often quite bulky and expensive, here is a simple, low-cost alternative.

DESCRIPTION

The circuit shown in Fig 1 is based around a two-transistor multivibrator, designed to produce square-wave oscillations at about 1kHz.

Harmonics (at 3kHz, 5kHz, etc) ensure the signal can be used not just to test audio circuits but RF circuits as well. When listening to the signal, it can best be described as a rather unpleasant buzz.

The multivibrator consists of a two-transistor circuit in which the transistors are alternately turned on and off. D1 and D2 are used to ensure the multivibrator produces square waves with very 'sharp' edges, and hence the greatest harmonic content. In theory, the harmonics from this circuit continue to infinity but in practice there is a limit at which they can be detected. With the circuit shown the harmonics are detectable beyond 145MHz.

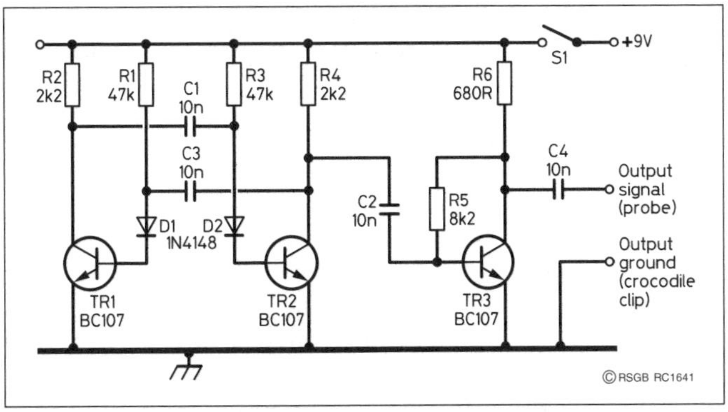

Output from the oscillator is coupled to an amplifier (TR3) via C2. The amplifier is used to ensure that any loading imposed by the circuit under test will not cause the multivibrator to stop. Biasing resistor R5 is low in value compared to many audio amplifiers but this ensures that the harmonic content of the output is as high as possible. Output coupling is via C4, which must be rated at 50V minimum to provide isolation and protection from the circuit under test.

Fig 1. The simple signal injector is based on a multivibrator. This generates square waves, which are rich in odd harmonics

CONSTRUCTION

The circuit is built on Veroboard (Fig 2), nine strips × 30 holes. There are several track cuts, which can be made with a special track cutting tool or a drill bit held in your hand. You can assemble the circuit in any order but generally the diodes and transistors are left to last. The output probe (Fig 3) is made from stiff copper wire, with the insulation left on for the majority of its length and the tip sharpened with a file to a point to ensure good contact. A lead with a crocodile clip is attached to the earth or chassis of the equipment under test.

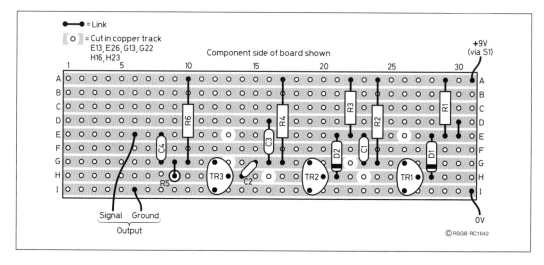

● ● = Link
[o] = Cut in copper track
E13, E26, G13, G22
H16, H23

Component side of board shown

+9V
(via S1)

Fig 2. Veroboard layout of the simple signal injector

TESTING AND USE OF THE SIGNAL INJECTOR

To test the circuit, touch the probe to either the input of an audio amplifier or the antenna socket of a receiver. If everything is working, you will hear a buzzing sound.

The basic idea of using a signal injector is to 'chase' backwards through a receiver, listening for the signal. By breaking a receiver into blocks, you can quickly isolate a fault to one block. Once located, you can pursue the fault to component level.

SAFETY NOTICE

The simple signal injector is designed to be connected to equipment which is powered on. Although there is a capacitor to provide protection from the voltages present in working circuits, the output must *not* be connected to any item of equipment which works on voltages higher than 24. This precludes television sets or any equipment which uses valves.

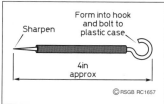

Sharpen

Form into hook and bolt to plastic case

4in approx

© RSGB RC1657

Fig 3. The output probe is made from stiff copper wire. Electrical connection is via a solder tag

Table 1. Components list

Resistors
R1, 3	47k
R2, 4	2k2
R5	8k2
R6	680R

All resistors ¼W, 10% tolerance, carbon or metal film

Capacitors
C1–3	10n 16V (minimum) disc ceramic
C4	10n 50V (minimum) disc ceramic

Semiconductors
D1, 2	1N4148 or any silicon signal diode
TR1–3	BC107/BC108/BC109 or any similar npn small-signal type

Additional items
S1	On-off switch

Veroboard (stripboard), 0.1in pitch, 9 strips × 30 holes
PP3 battery clip
Hook-up wire
Crocodile clip
Plastic case to suit

An amplified RF probe

When constructing radio-related projects, an RF probe is an item of equipment which is very useful to have around. It is also something simple to build. This RF probe employs a field effect transistor (FET) amplifier to increase its sensitivity.

BACKGROUND

RF probes are often built to be used in conjunction with a multimeter. This ties up the multimeter, which cannot then be used to make other measurements at the same time. This simple project, the circuit of which is shown in Fig 1, adds an FET amplifier to the basic probe. The whole project is housed in a small metal box, complete with 9V battery.

The meter required is not critical in nature. The prototype used a scrap item from a tape deck, which had a full-scale deflection (FSD) of 200µA.

CONSTRUCTION

Components are wired up point to point, as shown in Fig 2. The probe is soldered to the tag strip, the insulation on it keeping it clear of the metal case. The on-off switch S1 is part of a switched potentiometer (RV1), but you could just as easily use a toggle (or other) switch.

In use, attach the crocodile clip to the ground (earth) of the equipment you are checking

The probe itself is made from stiff copper wire. This needs to be covered with sleeving along most of its length, especially where it passes through the hole in the box.

Diode D1 can be any small, germanium, point-contact diode, but a Schottky type would make the unit more sensitive.

The photograph shows the connection of the short earth lead to a solder tag on the metal case. The crocodile clip on the end of this lead *must* be connected to the earth of the equipment being tested.

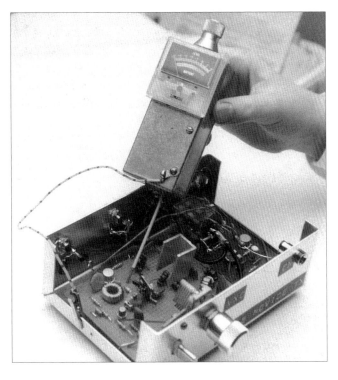

OPERATION

Switch the probe on, then adjust RV1 so that, with no signal present on the input, the needle of M1 is just above the zero mark. With the probe touching an RF signal source, you should see the meter

Right: Fig 1. The amplified RF probe uses D1 to rectify an RF signal, TR1 to amplify it, then M1 to display it

Below: Fig 2. Physical layout. Note the insulation on the probe where it passes through the metal box

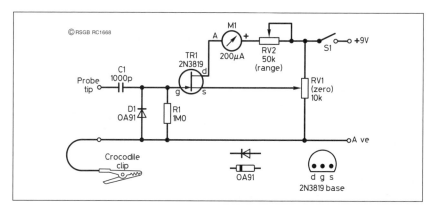

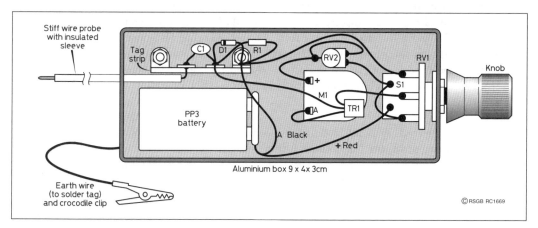

needle rise. RV2 is used to control the sensitivity of it. Although it would be possible to calibrate the RF probe against a known millivoltmeter, in most instances it is only required to either see whether a signal is present or not, or to adjust a circuit to peak a signal.

A wave trap for broadcast stations

Nearby medium-wave broadcast stations can sometimes cause interference to HF receivers over a broad range of frequencies. This being the case, set a trap to 'catch' the unwanted frequencies.

OPERATION

The way the circuit works is quite simple. Referring to Fig 1, you can see that it consists essentially of only two components, a coil L1 and a variable capacitor C1. This series-tuned circuit is connected in parallel with the antenna circuit of the receiver.

The characteristic of a series-tuned circuit is that the coil and capacitor have a very low impedance (resistance) to frequencies very close to the frequency to which the circuit is tuned. All other frequencies are almost unaffected. So, if the circuit is tuned to 1530kHz, for example, the signals from a broadcast station on that frequency will flow through the filter to earth, rather than go on into the receiver. All other frequencies will pass straight into the latter. In this way, any interference caused in the receiver by the station on 1530kHz is significantly reduced.

The wave trap can be roughly calibrated to indicate the frequency to which it is tuned

CONSTRUCTION

This is a series-tuned circuit which is adjustable from about 540kHz to 1600kHz. It is built into a metal box to shield it from other unwanted signals, and connected as shown in Fig 1.

To make the inductor, first make a 'former' by winding two layers of paper on the ferrite rod. Fix this in place with sticky tape. Next, lay one end of the wire for the coil on top of the former, leaving a few centimetres sticking out beyond the end of the ferrite rod. Use a couple of turns of sticky tape to secure the wire to the former. Now, wind the coil along the former, making sure the turns are in a single layer and close together. Leave a few centimetres of wire free at the end of the coil. Once again, use a couple of turns of sticky tape to secure the wire to the former. Finally, remove 1cm of enamel from each end of the wire.

Alternatively, if you have an old AM

Fig 1. The wave trap consists of a series tuned circuit, which 'shunts' signals on an unwanted frequency to ground

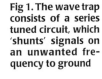

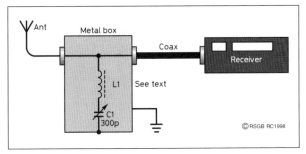

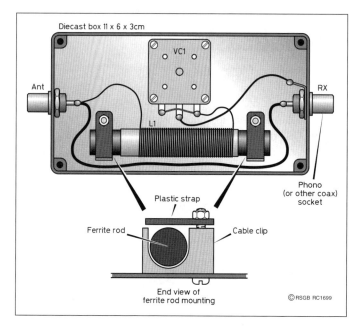

Diecast box 11 x 6 x 3cm

Table 1. Components list

Inductor
L1   80 turns of 30SWG
      enamelled wire, wound
      on a ferrite rod

Capacitor
C1   300p polyvaricon
      variable

Associated items
Case (die-cast box)
Knob to suit
Sockets to suit
Nuts and bolts
Plastic cable clips

Fig 2. Wiring of the wave trap. The ferrite rod is held in place with cable clips

transistor radio, a suitable coil can usually be recovered ready-wound on a ferrite rod. Some transistor radios will have coils for long wave and medium wave wound on the ferrite rod, in which case you have to make sure you select the right one. The medium-wave coil is usually a long, single-layer coil, while the long-wave coil is usually pile-wound. Ignore any small coupling coils.

Drill the box to take the components, then fit them in and solder together as shown in Fig 2. Make sure the lid of the box is fixed securely in place, or the wave trap's performance will be adversely affected by pick-up on the components.

CONNECTION AND ADJUSTMENT

Connect the wave trap between the antenna and the receiver, then tune C1 until the interference from the offending broadcast station is a minimum.

You may not be able to eliminate interference completely but this handy little device should reduce it enough to listen to the amateur bands.

I live near an AM transmitter on 1530kHz, and the signals break through on my top band (1.8MHz) receiver. By tuning the trap to 1530kHz, the problem is greatly reduced. If you have problems from more than one broadcast station, the problem needs a more complex solution.

A low-voltage alarm for battery supplies

Simple, direct-conversion receivers are prone to mains hum, and battery power prevents it. Assuming you use a rechargeable battery to power such a receiver, it should be recharged when the voltage falls below a predetermined level. Voltage checks can be made with a meter, but this can often be overlooked and the battery then becomes too flat to operate the equipment. This low-voltage alarm sounds when the voltage drops below a certain level to remind you that the battery needs to be recharged.

OPERATION

The alarm, shown in Fig 1, is connected across the receiver when in use. It is basically an oscillator circuit designed around a unijunction transistor TR1. The emitter of TR1 is biased by the resistor R1 and the Zener diode ZD1. The circuit will oscillate when the emitter has a sufficiently low voltage on it, and will cease oscillation if this voltage rises.

ZD1 must be chosen to determine the alarm voltage you require, and some figures for your guidance are given in Table 1. To obtain alarm voltages between the values given in the table, adopt the following procedure. Let us suppose you want an alarm voltage between 10.9 and 12.1V. Choose the Zener diode voltage that gives a 10.9V alarm, and insert a forward-biased silicon diode in series with the Zener, the cathode of the silicon diode being connected to the cathode of the Zener. This effectively increases the Zener voltage to 7.2V (ie 6.6V plus the 0.6V across the silicon diode). This will raise the effective alarm voltage to between 10.9 and 12.1V.

The alarm draws a small amount of current all the time so, although it could be left connected permanently across the battery itself, this is not recommended as it would eventually run it down.

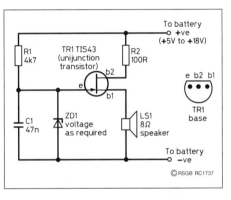

Fig 1. The low voltage alarm uses a unijunction transistor as an oscillator

Table 1. A guide to the alarm voltages produced by various Zener diodes	
Zener voltage	Alarm voltage
6.6V	10.9V
8.2V	12.1V
12V	19V

Table 2. Components list

Resistors
R1   4k7, ¼W, 5%
R2   100R, ¼W, 5%

Capacitors
C1   47n polyester

Semiconductors
ZD1   see text
TR1   TIS43

Additional items
LS1   8R, 1W
Case to suit (if required)
PCB, 2cm × 5cm

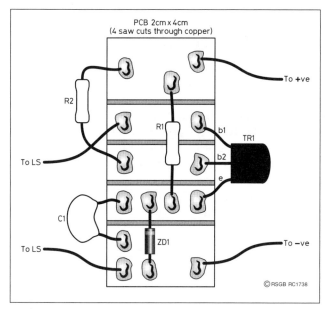

PCB 2cm x 4cm
(4 saw cuts through copper)

To +ve

R2

R1

b1

TR1

To LS

b2

e

C1

ZD1

To LS

To -ve

© RSGB RC1738

Above: **Fig 2. Physical layout. Before commencing assembly, make four saw-cuts through the copper of the PCB**

Above right: **The PCB can be fixed to the back of the loudspeaker with adhesive pads**

CONSTRUCTION

The circuit is built on a small piece of plain, single-sided PCB. Four saw-cuts are made through the copper, as shown in Fig 2, to make 'pads' to which the components are then soldered. When cutting, be careful not to cut right through the board, just the copper. I used a junior hacksaw to do this. The completed alarm can either be built into a case of its own – as I did – or into an item of equipment.

LIMITATION

The low voltage alarm relies on there being sufficient voltage in the battery you are monitoring to operate the alarm itself. Consequently, if the battery falls to such a low voltage that the oscillator will not operate, the alarm will not sound. Your receiver will not operate either, so you have two clues, both pointing to a flat battery!

A data interface

Interest in data communications such as packet radio, RTTY and PSK31 has always been high but, with the ready availability of software such as MMTTY, DigiPan and JVFAX, activity on other data modes is rapidly growing. The fact you can send and receive these modes from a PC with a sound card means that newcomers are more than willing to try them.

FIRST THINGS FIRST

The main problem that amateurs encounter is connecting a terminal node controller (TNC) or similar equipment directly to a hand-held transceiver. This is due to the way in which many hand-held transceivers use the transmit audio path to also control the transmit/receive switching (PTT) as well.

This interface, shown in Fig 1, is designed to connect a TNC directly to a transceiver, without having to modify either piece of equipment. Incidentally, I have used the term 'TNC' throughout this article, but it could just as easily be a different type of data interface. If a PC sound card is to be used, a slight modification will be necessary to the PTT circuit; the software being used will normally have circuit details in its help file.

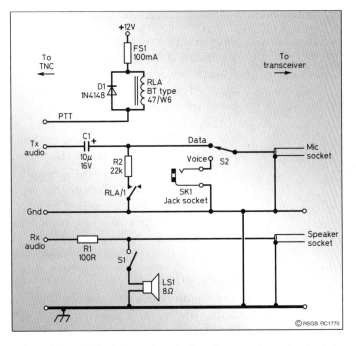

CIRCUIT OPERATION

The interface can be considered as having three main operational areas. These are:

1. The receive audio path.
2. The transmit audio path.
3. Switching.

The receive audio path is very simple, as it only comprises R1, which is designed to act as a protective load to prevent the audio amplifier in the transceiver being damaged in the event of a short-circuit on the lead between the interface and the TNC. LS1 is present, so that you can monitor the audio path if necessary but, to prevent operator fatigue when operating packet, it can be turned off by S1.

The transmit audio path uses S2 to select between data and voice. The voice connection is made via the jack socket SK1, which is provided so that

Fig 1. Circuit of the simple data interface for hand-held transceivers

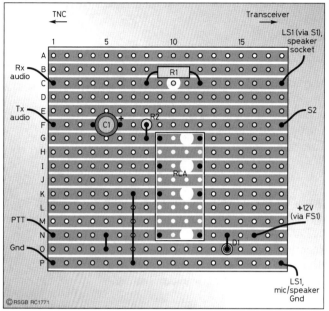

Table 1. Components list

Resistors
R1   100R, ¼W carbon
R2   22k, ¼W carbon

Capacitor
C1   10µ, 16V electrolytic

Semiconductor
D1   1N4148

Additional items
FS1   100mA fuse and holder
LS1   8R loudspeaker
S1   SPST
S2   SPDT
SK1   socket to suit external
      microphone

Relay, BT type 47/W6 (Maplin
DC80B)
Veroboard, 8 holes × 16 strips
Case, metal box (size to suit)
Hook-up wire

Fig 2. Veroboard layout. Note that D1 and R2 are mounted on their ends

an external microphone can be connected. C1 is used as a DC block between the TNC and the transceiver, so that the polarising voltage which is used in it to power electret microphones and control the PTT switching cannot damage the TNC. R2 and the relay contact RLA/1 are used to switch the transceiver from receive to transmit.

The switching path uses a small relay, which is powered from a 12V source, and the PTT switching contacts in the TNC activate it. D1 is used to suppress any back-EMF generated by the relay coil.

The only thing that needs to be considered is the transmit delay (TXDelay) parameter in the TNC. This might need to be lengthened to about 60ms to allow time for the transceiver to switch from receive to transmit. This shouldn't need to be adjusted, as the relay chosen has a fast switching time.

FS1 is used to protect the TNC and the interface in the event of a short-circuit. For portable operation, an alkaline PP3 battery could be used.

CONSTRUCTION

The prototype interface was built on Veroboard measuring 18 holes × 16 strips. The layout is shown in Fig 2. Alternatively, the interface could be built using point-to-point wiring. The only two components needing care when fitting are D1 and C1. Both are polarised components and need fitting the correct way round. The unit can be tested before being cased but, before using on a full-time basis, the interface should be housed in a metal box for screening purposes. The choice of connectors between the transceiver and the TNC will very much depend on the equipment in use. If the transceiver seems reluctant to switch from transmit to receive, you can try reducing the value of R2, but do not reduce this below 10kΩ.

An audio filter

This self-contained and inexpensive audio filter fits in the lead between a receiver and the headphones. Active audio filter circuits have been published in amateur radio journals over many years. Some are very elaborate and seem to have as many components as the receiver itself. Some are very clever, having several controls for adjusting the bandwidth and the frequency response, the notch or peak tuning, and other functions. These tend to invite the operator to spend more time testing and driving the filter than operating the receiver! Read on and find out why this one is different.

ABOUT THE FILTER

This audio filter unit is elementary in concept and has only one control – a switch to turn the battery on and off and at the same time switch the filter in and out of circuit. Thus, instant checks can be made on the effectiveness of the unit. In these days of DSP, the inexpensive, simple, single-integrated-circuit audio filter still has a useful place in amateur radio. It can perhaps best be described as an *audio pass-band modifier*.

The filter is a very effective and useful accessory for CW and SSB reception. To keep it simple and to avoid adjustable controls, a 'most-often-used' response characteristic has been adopted, with a peak around 800Hz. This frequency was chosen principally for CW reception but is also usefully placed and shaped for SSB reception. It attenuates both the low and the high audio frequencies, narrowing the pass-band.

This narrower pass-band reduces the unwanted noise heard in the headphones. However, this filter cannot eliminate or reduce the image sideband from the direct conversion process used in simple receivers.

THE CIRCUIT

The active element of the circuit, shown in Fig 1, is an operational amplifier integrated circuit, using the common, inexpensive, general-purpose LM741, with a bridged RC filter connected in its feedback path. How it works is a very interesting academic study which need not concern us here.

The RC filter comprises C2, C3 and C4, along with R4, R5, R6 and R7. The values of these filter components are important. Use good-quality components, preferably of identical type. The normal tolerance spread of component values results in a flat 'peak' in the response characteristic, making it acceptable for SSB and CW reception. Several filters have been built using store-bought components with the values shown. No special selection seems necessary. A 9V battery powers the unit. Remember to observe polarity and to switch it off when not in use.

CONSTRUCTION

The whole device fits into a plastic box. Select your own type! The double-pole changeover toggle switch is mounted on the lid of the box. The IC is

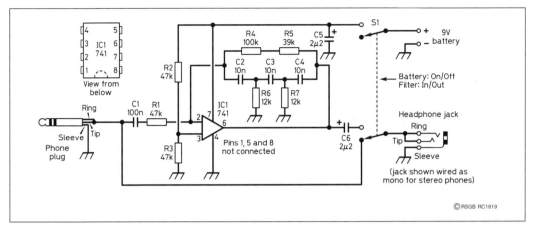

Fig 1. The simple audio filter. Note the pin assignment of IC1 shows the IC as it would be seen when wired-up 'dead bug' style

glued upside down to a piece of printed circuit board which is bolted to the floor of the box. Wiring is by the 'dead bug' method. Note that IC pin numbers 1, 5 and 8 are not connected. Two pieces of tag strip hold the RC filter components and support other wiring junctions. The battery is held in place in the box by double-sided sticky tape.

There is nothing critical about the construction, and instability was not experienced. The whole unit can be built in an evening. It has a high 'go-first-time' factor. There is nothing that needs any setting-up but feel free to experiment!

Table 1. Components list

Resistors

R1–3	47k, 5%
R4	100k, 1%
R5	39k, 1%
R6, 7	12k, 1%

All resistors ½W metal film

Capacitors

C1	100n
C2–4	10n, 1%
C5, 6	2μ2, 16V electrolytic

Semiconductor

| IC1 | LM741 |

Additional items

| S1 | DPDT toggle |

Battery clip
PP3 battery
Case
Audio connectors (as appropriate)

TESTING

To plot a filter frequency response you need access to quality laboratory instruments with known specifications. Trial tests using lesser-grade 'test meters' have been found to be prone to measurement errors, due in particular to insufficient information being known about the frequency characteristics of the measuring apparatus used. Consequently, no simple procedure for testing with cheaper test gear is offered. However, it is easy and convincing to check the effectiveness of the unit by ear.

Listen to a strong carrier signal, preferably with no modulation. Tune to what you estimate to be a typical signal tone or note for CW reception – about 800Hz. Switch the filter in and out of circuit. Adjust the pitch of the received signal very carefully. You should find a position where the amplitude (loudness) of the wanted signal rises when the filter is in circuit. Ignore any change in the noise level, just listen to and refer to the change in the level of the received signal alone, as the filter is switched in and out of circuit. Above and below this frequency you should find that the loudness of the signal is higher when the filter is out of circuit, confirming a peak in the overall filter response. There is a very noticeable reduction in noise when the filter is in circuit.

A 'loop' alarm

Putting on a station at a local gala or similar function is a good way of gaining publicity for our wonderful hobby but, when you are on a stand, it is not always possible to keep an eye on the equipment. This alarm is very useful for those situations. It is similar to those used in shops, where a loop of wire is wrapped around the goods. If the wire is broken to remove an item, the alarm sounds.

HOW IT WORKS

The alarm uses a thyristor, which is also known as a *silicon controlled rectifier* (SCR). [To avoid any misconception by reading the name wrongly, this is a *rectifier* that can be *controlled*, and it is made out of *silicon – Ed.*] An SCR is like any other diode in that it will only allow current to flow in one direction, ie from the anode, 'a', to the cathode, 'k'. However, unlike any other diode, this current will only start to flow when a small positive voltage is applied to the gate, 'g'. Having started to conduct, the SCR continues to do so, even if the gate voltage is removed.

In the circuit, shown in Fig 1, the SCR does not conduct under normal conditions because the wire loop maintains zero volts on the gate. If the loop is broken, the gate is pulled positive by R1 and R2, which makes the SCR conduct, sounding the alarm. Even if the loop is re-joined, the SCR continues to conduct and the alarm continues to sound until the power is turned off.

Above: The completed project. In this photo the keyswitch cannot be seen, as it is on the far side of the box

Below: Fig 1. The loop security alarm uses the wire loop to prevent the SCR from being turned on

CONSTRUCTION

The circuit is built on a small piece of single-sided PCB. Saw cuts are made through the copper, as shown in Fig 2, to form pads. The components are soldered to these pads. The on/off switch

Table 1. Components list

Resistors
R1   10k
R2   2k2
R3   470R
All resistors ¼W

Capacitors
C1   100n
C2   100µ electrolytic

Semiconductor
TH1   C106 (Maplin)

Additional items
Key-switch
Piezo buzzer (Maplin)
Plastic case
PCB material
Wire for loop
Plug(s) and socket(s) for loop

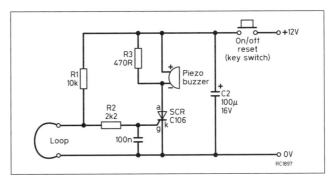

Fig 2. Construction is simple on a small piece of PCB, but be careful when cutting through the copper not to cut right through the board

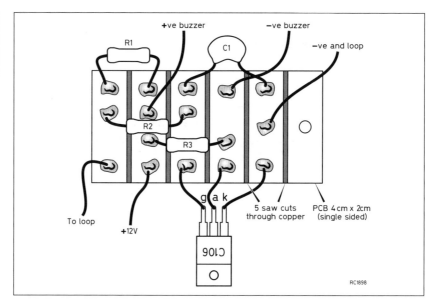

should ideally be a key-switch, to prevent the alarm being switched off by an unauthorised person.

The ends of the loop can be connected via practically any type of connector to the case containing the alarm; indeed, the loop could comprise of any number of short lengths of wire, joined with in-line plugs and sockets. The latter would allow a single item of equipment to be removed without disturbing the whole loop.

A mobile microphone

Many radio amateurs enjoy operating mobile, and there are many times when having a radio in a car can, in fact, help make a long journey seem to take a shorter time. Unfortunately, problems can occur if you use a hand-held microphone. Not only can you cause an accident if the cable interferes with the controls of your car, you are also in breach of the Highway Code. This hands-free microphone was designed so that you can operate while still being in full control of your vehicle.

CONSIDERATIONS

There are two aspects which must be considered:

- the microphone and its fitting;
- transmit/receive switching.

The circuit for the microphone is shown in Fig 1. The microphone chosen for this project is an inexpensive electret type, available from many sources, including rallies and many of the major electronic component suppliers. Electret microphones are ideal for this project since they have a reasonably high audio output level, small size, and high sensitivity. They do, however, require a power feed of between 1.5 to 9V at about 5µA to power the microphone's integral amplifier. There are basically two types of electret capsule. The first has three connections, ie ground, audio and DC power. It is not really suitable for this circuit. The second type has the DC power and the audio combined. This means that only a single screened lead is needed to connect the microphone assembly to the switch box.

Fig 1. Circuit of the mobile microphone. Use of an electret capsule which uses just two connections means that single screened cable can be used between the microphone assembly and the switch box

Table 1. Components list

Resistor
R1   1k, ¼W

Capacitors
C1   10µ electrolytic
C2, 3   10n ceramic

Additional items
S1   DPST toggle switch

Battery clip
PP3 battery
PP3 box to suit
Screened cable
Microphone (Maplin FS43 or similar)
Plug for microphone (to suit transceiver)
3.5mm in-line plug and socket

POWER

In many modern transceivers, a voltage is available on the microphone socket to provide power to external microphones, but care must be taken when using this supply as it is often not fused. The DC

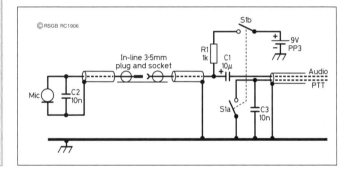

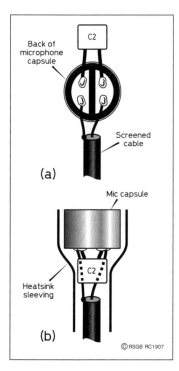

Fig 2. Solder C2 and the screened cable to the back of the microphone insert (a), then cover the assembly in heat-shrink sleeving or self amalgamating tape (b)

power feed to the transceiver from the car can also contain a lot of interference. Consequently, a 9V PP3 battery was used for the power source. With the low current required, the battery should last for years. R1 is used to act as a current limit, to prevent excessive current flow in the event of a short circuit. C1 is used as a DC block to prevent damage being caused to the transceiver by the DC supply. C2 and C3 are used to provide RF decoupling, to prevent any RF pick-up from the transmitter getting into the audio.

SWITCHING

Transmit-receive switching can be achieved in many ways. I have tried various ideas, such as timed circuits or press-to-transmit, press-to-receive circuits, but they have all had problems. For simplicity, I have always returned to a simple switch. A voice-operated system was discounted due to the fact that comments made by passengers could be transmitted.

CONSTRUCTION

The microphone assembly is very simple, with the microphone capsule being soldered directly to the cable with capacitor C2, then covered in heat-shrink sleeving or self-amalgamating tape. The connections to the microphone are shown in Fig 2. A small piece of foam or cloth can be fixed over the microphone capsule if required to reduce the effects of wind and breathing noise.

The microphone can then be fitted in a number of ways. I provide below two ideas as a basis for experimentation:

- Fix the microphone assembly to a crocodile clip, as shown in Fig 3, so that it can be clipped to your lapel or (assuming you wear them) the stem of a pair of spectacles.
- Fix the microphone assembly to a stiff piece of copper wire to form a head or neck band.

It is a good idea to include a 3.5mm in-line jack plug and socket on the microphone cable so that, if you get out of the vehicle without taking the microphone off, the lead will disconnect without damaging anything.

It is advisable not to fix the microphone to the car and have it sticking out in front of your mouth because, in the event of sudden braking, you may find yourself being thrown forward onto it.

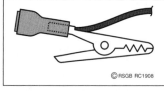

Fig 3. How the microphone assembly can be fixed to a crocodile clip for use as a lapel microphone

The switching box is used to hold the battery, R1, C1, C3, and S1. As the construction of the circuit is very simple, I leave the arrangement of the components up to you but you could leave sufficient room inside the box to mount either a tone-burst or up/down tuning buttons. This box should be positioned where the switch is easy to operate but clear of the controls of the car.

An audio power booster

When operating from the car, most radio amateurs have experienced, at one time or another, difficulty in hearing a distant station. This may be due to a noisy old car, a chatty girlfriend, or the occupant of the next car who insists on playing his/her music too loud! Whatever the reason, I wanted to hear more. I had tried connecting an external loudspeaker, but unfortunately most radios only produce about 1W of audio power. This is not sufficient to drive a loudspeaker well without distortion. To counteract this problem I decided to design a booster amplifier for my radio, thereby increasing the level of audio to about 10W to an external speaker. Since installing the booster in the car I have been able to hear even the quietest or far away stations.

HOW IT WORKS

The booster (circuit diagram in Fig 1) is based around the TDA2005M audio amplifier chip. This was selected because it delivers high output power from a 12V supply, as found in a car. Most other chips require a far greater supply voltage and are unsuitable for this application. High output power from a low supply voltage is obtained by operating in a bridge mode. The loudspeaker is connected between the outputs of *two* power amplifiers, operated in opposition. Care must be taken not to earth either loudspeaker connection, otherwise the chip will be damaged.

One point of interest is the inclusion of an RF filter (R1, C1 and C2). This helps prevent RF pick-up from the audio lead running from the radio to the booster. Also included is an input level adjustment, VR1, which allows the output level of your radio to be matched to the input of the booster.

HOW IT IS BUILT

Because of the odd pitch of the legs of the TDA2005M, it is not possible to construct this circuit using stripboard without a lot of surgery to the chip.

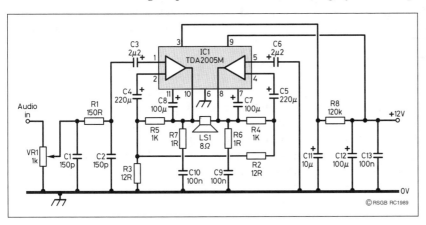

Fig 1. The audio power booster uses a TDA2005M, which can produce several watts of power comfortably from a 12V supply

© RSGB RC1989

47

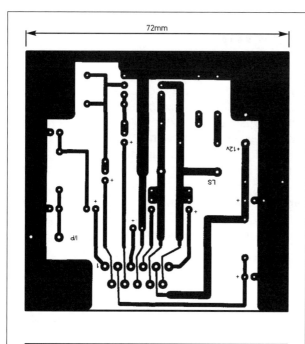

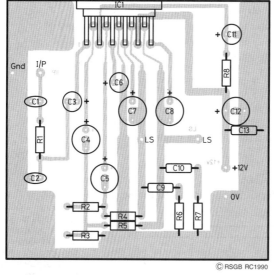

© RSGB RC1990

Table 1. Components list

Resistors

R1	150R
R2, 3	12R
R4, 5	1k
R6, 7	1R
R8	120k
VR1	1k logarithmic potentiometer

All resistors 5% 0.25W metal film

Capacitors

C1, 2	150p ceramic plate
C3, 6	2µ2, 25V radial electrolytic
C4, 5	220µ, 25V radial electrolytic
C7, 8, 12	100µ, 25V radial electrolytic
C9, 10, 13	100n metallised polyester film
C11	10µ, 25V radial electrolytic

Semiconductor

IC1	TDA2005M

Additional items

10W 8R loudspeaker
Vero pins
Heatsink
PCB
Wire
Connectors

Fig 2. (a) PCB foil pattern, and (b) component placement for the audio power booster

Therefore, I designed a PCB (Fig 2(a)) to facilitate the building of the project. You can use whatever mode of construction appeals to you. Take care not to make solder bridges between the tracks, particularly around the power amplifier chip where the tracks are close together.

When constructing the amplifier, first solder the resistors, followed by the capacitors, and finally the chip. When installing the electrolytic capacitors, pay attention to the correct polarity, as shown in the component placement (Fig 2(b)). In the prototype, no provision was made to mount VR1 on the PCB, as it was envisaged that most people would wish to use a panel-mounted potentiometer.

It is recommended that the amplifier be built in a metal box. Additionally, you may find it necessary to wrap the audio cable and the power cables around a ferrite ring, to prevent RF feedback. The heatsink tab of the audio amplifier is earthed and may be connected directly to the metal box to act as a heatsink.

SETTING UP

After checking the PCB for dry joints, short-circuits and correct polarity of the capacitors, connect the amplifier to a 12V power supply via a 2A fast-blow fuse. Adjust VR1 fully anticlockwise and connect the amplifier to your radio. Switch on the power supply. Adjust the volume control on your radio to approximately half way, then adjust VR1 to a comfortable maximum volume, free of distortion. You may then adjust the audio level either from the radio's volume control or VR1 if it is panel-mounted.

A portable power supply

When time permits, I enjoy participating in the Backpacker series of RSGB contests [1], operating in the 3W category. The power supply I use for these outings is described here. It is ideal for the purpose, and can also be used as a simple uninterruptible power supply (UPS) for the shack.

DESIGN CRITERIA

- A maximum of 4.5kg (10lb) in weight.
- Able to be carried in a small rucksack.
- Able to be plugged into the nearest mains outlet to be recharged.
- When at home to run in 'float-charge mode', to operate low-current equipment.
- To be of reasonable cost.
- To use readily available components.

At the 1998 Rainham Rally I found a couple of sealed lead-acid cells rated at a nominal 12V and 7Ah capacity. They weighed in at a little over 2.2kg (5lb) – just what I needed!

Being of the sealed variety, care has to be exercised in not overcharging the cell, ie to prevent gassing, so 13.8V would be the maximum permitted voltage at the terminals. Thus a stabilised supply was essential.

THE CIRCUIT

The completed power supply

The float charger is hardly original – indeed it was adapted from that excellent series of articles by John Case [2]. One major consideration was the need to 'over-engineer' it, as I would leave it plugged in and switched on almost continuously.

Fig 1 shows the circuit diagram. Many of the components were salvaged from redundant equipment or the junk box; however, the critical components – transformer, pass transistor, reservoir capacitor and regulator chip – were all purchased new.

There are many options available for layout; mine were dictated by the size and shape of the heatsink. In my case, the box that holds it all is made from 12mm plywood, with the major components mounted on an L-shaped aluminium plate which forms the front and part of one side. A voltage-dependent resistor (VDR) is fitted across the primary of the transformer, and an over-current control is provided to limit the current to 1.5A.

Table 1. Components list

Resistors

R1	1k2
R2	1k5
R3	2k7
R4	0R5, 2W
R5	8k2
R6	7k5
R7	820R
RV1	500R linear preset potentiometer
RV2	1k linear preset potentiometer
VDR1	V275LA40A

All resistors ½W metal film, 5% tolerance, unless otherwise stated

Capacitors

C1	10,000µ electrolytic, Maplin LE03
C2	4µ7 electrolytic
C3	500p ceramic

Semiconductors

BR1	KBU4D (or similar)
D1	Red LED
D2	Yellow LED
D3	Green LED
D4	MR752 (or similar)
IC1	LM723
TR1	2N3055

Additional items

F1	1A plus holder
F2	3A plus holder
S1	Double pole, single throw toggle
T1	Mains transformer with 2 × 15V @ 0.75A secondaries (Maplin DH27)

7Ah sealed lead-acid battery
IEC socket
Matrix board
Screw terminals, insulated
Case to suit

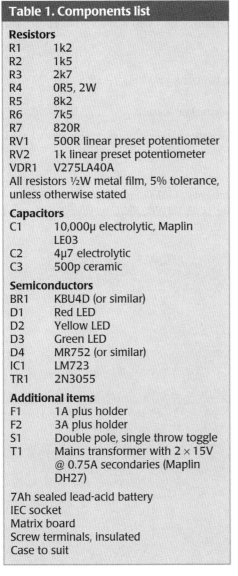

Fig 1. The portable power-pack works by float-charging a sealed lead-acid battery, to provide an uninterruptible supply

A heavy-duty diode D4 is incorporated in the feed to the battery, and LEDs are fitted at strategic points as a confidence feature and for ease of fault-finding. The voltage controller, an LM723, is fitted, along with its components, on a small piece of matrix board, the remaining components being wired 'point-to-point' using substantial cable for the heavy-current paths.

Setting-up is quite straightforward – connect everything together to a fully charged battery, connect a voltmeter to the battery's terminals, turn on and adjust the output to 13.8V as measured at the battery terminals. Set the current limit to 1.5A, which in practice appears rarely achieved.

RESULTS

Does it meet the criteria I set out earlier? That I will leave you to judge. The figures are:

- Weight – 11lb.
- Size – 25 × 25 × 10cm, including heatsink and handle.
- Cost – around £20.
- Capacity – enough for a full Backpacker session

Even when left on continuously, the temperature of the portable power pack hardly rises above room temperature.

REFERENCES

[1] 'Backpacking – summertime delights', G6TTL, *RadCom* May 1997.
[2] 'Power supplies on a shoestring', GW4HWR, *RadCom* July and August 1986.

Optical communication

Many principles of radio communications can be demonstrated very effectively by a wireless link using visible light. This is not surprising, since both light and radio waves are types of electromagnetic radiation, but light allows us to see what is actually happening. The system described here is a simple optical transceiver which has been a successful project for students and teachers in radio clubs at a number of local schools.

DESCRIPTION

The project uses an LED and photodiode, and is based around an LM324 'quad' operational amplifier (ie four op-amps in one package). The circuit diagram is shown in Fig 1 and the PCB foil pattern in Fig 2. The transmitter uses one of the op-amps, the receiver uses two, and there is one spare. Op-amps amplify any voltage difference between an inverting and non-inverting input, marked '−' and '+' respectively. In this system, the non-inverting input ('+') is used as a reference point at half the supply voltage, so any signal on the other input is amplified in comparison to this reference voltage. Amplifier gain is determined by the ratio of value of the feedback resistance to the input resistance (eg R1/RV1).

The preferred LED produces a very bright light with a narrow beam. Other LEDs may be significantly less powerful and will have more limited ranges. It is possible to use a photocell from a model kit as the detector, but the photodiode and LED can be obtained from a good component supplier (for example, Farnell Components, www.farnell.com). R10, R11, R12, C4 and C6 are included to ensure supply voltage stability.

In the transmitter, the LED glows continuously, since the transistor is

Fig 1. The low component count of the optical transceiver is largely due to the use of the LM324, a quad op-amp

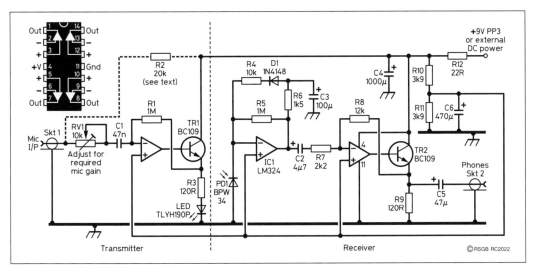

Transmitter   Receiver   ©RSGB RC2022

53

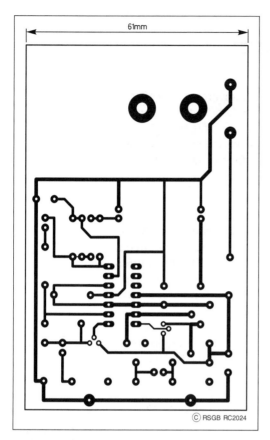

Fig 2. Foil pattern of
the single-sided PCB

61mm

© RSGB RC2024

turned half on by the op-amp. When a small signal is produced by the microphone, it is amplified by the op-amp which tries to turn the transistor on and off, making the LED flicker in sympathy with the input signal. Hence we have amplitude-modulated visible light. If you use a dynamic microphone the pull-up resistor R2 *must be* omitted but if you use an electret microphone it *must be* included.

The receiver begins with a photodiode, which varies in resistance according to how much light is present. Changes in resistance are amplified by an op-amp, but since this is still quite a small signal, a second op-amp is used to amplify the signal further. Again, a transistor is turned half on by the final op-amp, driving headphones as the audio output.

TESTING

The LED should glow as soon as the unit is turned on and, if testing takes place in the presence of domestic mains lighting, a loud hum will be heard because AC lighting 'flickers' at 100Hz and presents a strong signal to the receiver. TV screens, monitors and other electronic displays will produce a variety of sounds, according to their display refresh rates. Check the orientation of the LED if it is unlit, and similarly the photodiode if no signals can be heard. The unit should draw about 45mA, so anything less suggests incorrect assembly, while significantly more may indicate a short-circuit. The unit can be tested in isolation using a mirror but the light beam must hit the photodiode directly since a near miss will not produce an audible signal.

EVALUATION AND EXPERIMENTATION

The optical transceiver can be used to demonstrate many aspects of radio theory.

Amplitude modulation (AM)

Although the LED is modulated by an audio-frequency signal from the microphone, the light is only ever seen to flicker if distortion occurs by over-modulation. A more effective demonstration of AM is achieved by interrupting a beam of light with a fan, rather than modulating it at source. Any light source will do and although an electric fan can be used, a cardboard propeller on the end of a hand drill illustrates the principle most effectively.

Automatic gain control (AGC)

Radio receivers often include an automatic gain control, which reduces their sensitivity when strong signals are present. This receiver circuit includes a very simple automatic gain control, based on D1. If the voltage across the feedback resistor is less than 0.6V, this is insufficient to forward bias the diode, and feedback through the 1MΩ resistor R5 assures high gain. If the voltage across the feedback resistor exceeds 0.6V, the diode becomes forward biased and the resistance of the feedback loop becomes less than 10kΩ, reducing the gain significantly. The circuit will work without this feature, but is easily silenced by strong background light, which can be demonstrated by temporarily removing D1.

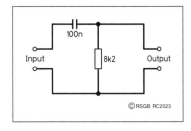

Fig 3. Simple high pass audio filter to reduce mains hum

Selectivity

Selectivity is the ability of a receiver to select one signal in preference to others. Although the photodiode is most sensitive to orange light, it will also detect light of other frequencies, which are visible as other colours. In fact, it will even detect infra-red signals invisible to the naked eye, so the unit can 'hear' the commands of a TV remote controller. The output of the Toshiba LED has significant infra-red content, so infra-red filters can be included to create an invisible link. Simple filters using coloured cellophane can be used to experiment with a selection of signals from different coloured lights. Offcuts from lighting filters used in local theatres may be obtainable from the stage lighting manager.

Noise filtering

It is evident from the background hum of the receiver in the presence of AC lighting that AM signals are susceptible to interference from other signals. In this case a simple 200Hz high-pass filter in series with the headphone output can be used to attenuate the low-frequency mains hum without blocking the wanted audio signal. This can be achieved using a simple RC filter, as shown in Fig 3. The nominal cut-off frequency is given by the formula below, and the effect on the signals heard is quite noticeable.

$$f = \frac{1}{2\pi RC}$$

DX CONTACTS

Power and antennas are the major considerations for DX enthusiasts. The LED specified is actually rated at 50mA, so the value of R3 can safely be reduced to 39Ω to increase power output (R12 then needs to be ¼W). Both receiver and transmitter performance can be significantly improved by the use of lenses or a magnifying glass. This focuses the signal carrier, just as directional antennas do in radio communications. A more focused beam can be obtained if the cap of

Table 1. Components list

Resistors

R1, 5	1M
R2	20k
R3, 9	120R
R4	10k
R6	1k5
R7	2k2
R8	12k
R10, 11	3k9
R12	22R
RV1	10k preset potentiometer

All resistors 0.1W 5% tolerance, unless otherwise stated

Capacitors

C1	47n
C2	4µ7, 16V
C3	100µ, 16V
C4	1000µ, 16V
C5	47µ, 16V
C6	470µ, 16V

Semiconductors

IC1	LM324
TR1, 2	BC109
D1	1N4148
LED	TLYH190P (Toshiba)
PD1	BPW34 photodiode

Additional items

Battery clip
Jack socket(s)

the LED is cut off (eg with a junior hacksaw) and the LED then polished on an abrasive stone and finished-off with Brasso on newspaper. The LED can then be mounted at the focal point of a lens. Do not shine the LED directly into anyone's eyes.

Experiments have achieved distances approaching 100m, and the authors would be keen to hear from anyone who betters this distance.

FURTHER INVESTIGATIONS

If the photodiode is connected to a crystal earphone with no other components, it will detect strong signals such as neon or fluorescent lighting in close proximity. Like a simple crystal radio receiver, no batteries are required, but lenses will improve performance significantly, just as a good antenna will for a radio receiver.

Other ideas could be to build a repeater, by taking the receiver output directly to the transmitter input, or to use a comparator interface to allow data and SSTV to be sent from a computer. The system can be used as a dedicated audio link or as an educational aid, but in either case it provides an excellent opportunity to understand some of the principles of telecommunications.

A diode/transistor tester

One of the regular problems when assembling diodes into a project is working out which end is the cathode and which the anode. This is particularly so when using diodes from the 'junk box' and finding that the markings have become somewhat indistinct. It is not too difficult to use a resistance meter to check the polarity but this only looks at one aspect. The unit described below provides an unambiguous means of identifying the cathode and carries out some other simple tests. It can also be used to check that bipolar transistors are 'in the land of the living'.

CAUTION

But first, a few words of caution: be very wary of using rectifiers of unknown characteristics in power supplies or other high-current applications. The fact that a rectifier looks big enough is not a good indication that it is adequate for the job. Using inadequate devices is likely to result, at best, in poor reliability. At worst, the result could be the well-known 'dark brown smell', followed by smoke and even flames.

CIRCUIT DESCRIPTION

The circuit is shown in Fig 1. A simple square-wave oscillator is formed by inverters IC1a and IC1b. This runs at about 1.2kHz although the actual frequency is not particularly important.

Gates IC1c and IC1d buffer the output of the oscillator and split the signal into two paths. One of the paths is then inverted, which results in two anti-phase square-wave signals being produced at the outputs of IC1c and IC1e. These anti-phase signals are fed to the inverting inputs of a dual operational power amplifier IC2 via R3 and R4. The non-inverting inputs are connected to a potential divider chain which holds them at half the supply voltage.

Consider, now, the diode under test (DUT) being connected to the test terminals TC1 and TC2 as shown. When IC2a output is high, current will flow through R5, D1, the DUT, LED2 and R6 to the output of IC2b, which will be at about ground potential. The result is that LED2 will illuminate.

The completed diode/transistor tester

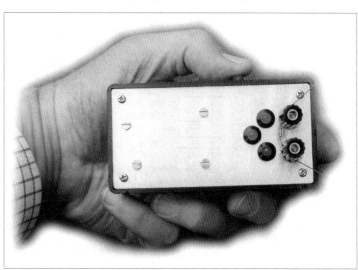

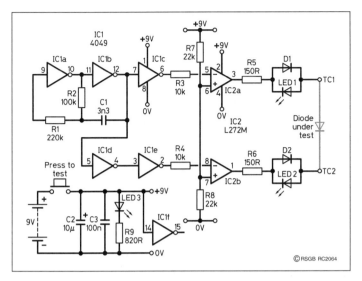

During the next half of the cycle, the output of IC2a will be at about ground potential and that of IC2b will be high. Under these conditions, the DUT will be reverse-biased so no current will flow.

With the DUT connections reversed, it follows that the operation of the circuit will be similar to that previously described, but LED1 will now be illuminated instead of LED2.

If the unit is constructed so that LED1 is adjacent to TP1 and LED2 to TP2, a direct indication is given of the termi-

Fig 1. The simple diode tester gives an instant 'dead' or 'alive' indication

nal to which the cathode of the DUT is connected.

CONSTRUCTION

With the exception of the placement of C2 and C3, there is nothing at all fussy about the construction and the builder can use PCB, stripboard or

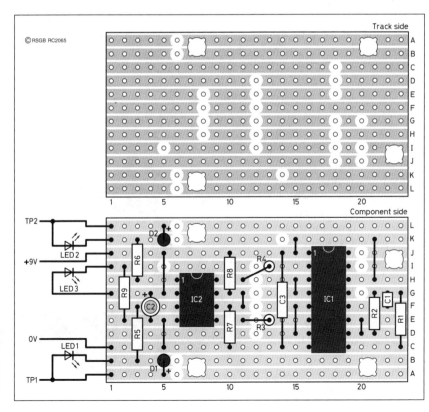

Fig 2. Stripboard layout and component overlay. Note that if you use the recommended project box, there is little room to spare

'dead bug' construction as preferred. Whatever form of construction is used, C2 needs to be connected as close as possible to pins 2 and 4 of IC2, and C3 to pins 1 and 8 of IC1.

A suitable stripboard layout is given in Fig 2. Some care is needed if the SB1A box is to be used. The width of the stripboard is such that it fits between the lid support pillars of the box with little room to spare. Check the placement of the board before drilling the mounting holes.

USING THE TESTER

Connect the diode under test between TP1 and TP2 and press the test button. You have a working device if either LED1 or LED2 illuminates, the illuminated LED indicating the cathode of the device.

Should both LEDs illuminate at equal brilliance, the diode is short-circuit or is a low-voltage Zener. If both LEDs illuminate but one is brighter than the other, the chances are that the device being tested is a Zener with a breakdown voltage in the 4.7–6.8V region. In this case, the brighter of the two LEDs indicates the cathode.

If neither of the LEDs illuminates, this indicates either an open-circuit diode or a flat battery – check that LED3 illuminates to eliminate the latter possibility.

Testing transistors takes a few more steps – see Fig 3. Steps 1 and 2 can be skipped initially and carried out if needed to give information on the probable fault should step 4 fail.

At step 1 it might be found that both LED1 and LED2 illuminate, with one being brighter than the other. This is indicating that the reverse base-emitter breakdown voltage is less than about 7V, which is normal for some transistors. In this case the brighter of the two LEDs is the one to note.

The transistor test is not particularly exhaustive but it can be a simple way of checking that a device is still 'alive'.

Table 1. Components list	
Resistors	
R1	220k
R2	100k
R3, 4	10k
R5, 6	150R
R7, 8	22k
R9	820R
All resistors metal film ¼W, 5%	
Capacitors	
C1	3n3 ceramic (eg Maplin RA41U)
C2	10µ, 16V tantalum bead (eg Maplin WW68Y)
C3	100n polyester (eg Maplin BX76H)
Semiconductors	
IC1	4049
IC2	L272M
D1, 2	1N4001
LED1–3	5mm high-brightness (eg Maplin WL84F)
Additional items	
Push button switch, eg Maplin FH59P	
4mm terminal posts, eg Maplin FD69A	
Project box, eg SB1A – Maplin BZ27E	
Stripboard	

Fig 3. Use of the device to test bipolar transistors

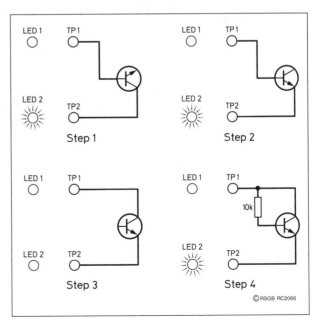

Step 1

Step 2

Step 3

Step 4

© RSGB RC2066

An HF antenna with gain

Although we may have passed the current sunspot peak, activity on the HF bands is still very high. One disadvantage that many new operators have is that they are limited to just a simple dipole antenna and perhaps low power. Described here is a simple wire antenna which has a gain of about 3dB over a dipole, but has the advantage over a beam in that it is simple to construct and more particularly is as unobtrusive as a dipole. No originality is claimed for the antenna, which is based on the 'Zeppelin' antenna used on German airships up to the late 'thirties. This version is a stretched or extended double Zeppelin, which consists of two collinear elements fed in phase to give gain.

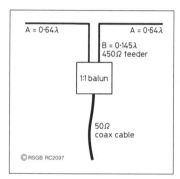

Fig 1. Schematic of the Zepp, with dimensions in wavelengths. See Table 1 for the dimensions for various amateur bands

PROPORTIONS

The antenna consists of two elements 0.64λ long, which are centre-fed using a section of 450Ω open-wire feeder. The feeder could be extended all the way to the shack and a balanced ATU used to match the antenna to the transceiver but, instead of this and for convenience, a short section of 450Ω feeder is used to act as a matching section. The antenna is then directly fed with 50Ω coaxial cable ('coax').

The matching section is 0.145λ long but, as this is constructed from twin feeder, the velocity factor needs to be taken into account.

The antenna is balanced and coax is unbalanced so a 1:1 balun is used to reduce the possibility of interference. Such a balun can be constructed in a number of ways but, for cheapness, a coaxial choke balun can be used. This consists simply of five turns of coaxial cable, coiled in a 10cm diameter coil, located close to the feed point of the antenna.

Although the antenna was originally intended for 10m, the dimensions in Table 1 give the sizes for all bands from 15m to 2m. The prototype was built for 2m and all subsequent dimensions have been produced by scaling.

Table 1. Antenna dimensions

Band (m)	Frequency (MHz)	A length (m)	B length (m)
15	21.225	9.05	1.95
12	24.950	7.69	1.66
10	28.300	6.78	1.46
SSB 6	50.200	3.82	0.82
FM 6	51.500	3.73	0.80
4	70.250	2.73	0.59
2	145.000	1.32	0.29

CONSTRUCTION

The prototype antenna was made from 16SWG uninsulated copper wire but a better material for the HF bands would be a stranded wire such as Flex Weave as this would allow for some movement of the antenna in the wind. The top section should be cut to the dimensions shown in Table 1, and care taken to ensure that this dimension is from the electrical end of the wire, ie if an end insulator is used then this would be the centre point in the insulator and not the section which is folded back to provide the strength (Fig 2).

The size of the top sections should be rigidly adhered to, while the matching section of 450Ω feeder should be made slightly longer than given in the table so that the connection point to the coax can be adjusted to provide a good match as measured on a SWR meter. Care should be taken when measuring the lengths, especially on the VHF bands, remembering the old adage of measuring twice and cutting once. The 450Ω feeder used for the prototype was constructed from the same 16SWG wire as the main part of the antenna and the wires kept 25mm apart using pieces of hard plastic as spacers. This method of construction is suitable for VHF antennas as the length of the feeder is relatively short, but for the HF bands the use of commercial 450Ω feeder is a more practical option due to the extra length involved.

When constructing the antenna, care should be taken to ensure that all the joints are soldered and waterproofed to prevent the ingress of moisture, especially into the coaxial cable. The method of waterproofing I prefer is to encase the connections in a 35mm film container, and then seal it with self-amalgamating tape to ensure that it is watertight (see Fig 3). If you are not a photographer then asking at your local chemist or a shop with in-house film processing will often lead to a good supply of these containers.

An alternative container can be made from the plastic 'corks' which are used with some sparkling drinks. There are several sealing or caulking compounds available which should *not* be used, as they give off acetic acid (vinegar) fumes when curing and cause corrosion of the soldered joints and cable.

MOUNTING

The antenna can be mounted vertically or horizontally, although this will usually depend on the bands and modes used.

On the HF bands it is more conventional to use the antenna horizontally (horizontal polarisation), while on VHF the antenna can be mounted horizontally (horizontal polarisation) for SSB or vertically (vertical polarisation) for FM and packet.

The prototype antenna was made for 145MHz and, although it did not give as good a result at ground level as my main station antenna (a collinear at 8m), it showed a significant improvement on the flexible antenna supplied with my hand-held transceiver. As a result I could access a repeater about 45km from my house running 1.5W from a hand-held transceiver. The measured SWR was 1.3:1 at the band edges and 1.2:1 at the band centre.

On the HF bands the antenna should give a gain of about 3dB gain compared to a dipole, with the main direction of radiation at right-angles to the antenna (ie a north-south antenna will radiate east-west), although this will depend on the height above ground and nearby objects.

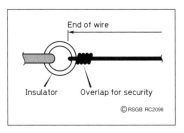

End of wire

Insulator   Overlap for security

© RSGB RC2098

Fig 2. The ends of the antenna wire are wrapped around insulators and soldered

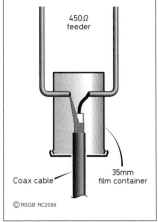

450Ω feeder

Coax cable

35mm film container

© RSGB RC2099

Fig 3. Use of a 35mm film canister to provide weatherproofing at the feed point

An auto-keyer

Here's a simple idea for those who enjoy improvisation and perhaps don't want to bother with keyer ICs or computers. Make an audio tone recording of your CW keying and use it to make repetitive CQ calls when band conditions are marginal. Assuming that a discarded or unused cassette recorder is to hand, this method is economically effective compared to more expensive keyers, which still cannot easily reproduce your own exact keying characteristic (do remember, though, that this is a 'warts and all' situation, and your keying could serve either to encourage or warn off potential contacts). To ensure the former, happy condition, make sure your call is easily readable and without any off-putting quirks.

TAPE RECORDER
Choose a mains/battery type that can accommodate a small PCB or scrap of stripboard inside its battery compartment – 6V to 9V versions are fine.

As it is necessary to locate three connection points inside the machine, a word of caution is appropriate regarding the danger of electrocution while working with the covers off. A low-voltage bench power supply or battery is highly recommended as a temporary power source for the machine while working on the innards!

DESCRIPTION
The block diagram is shown in Fig 1 and the circuit diagram in Fig 2. The audio frequency keying, a tone at a nominal 1kHz, is rectified by D1, then filtered and integrated by R1, R2, C1, and C2 to provide a positive-going DC voltage to the base of TR1, which switches on RLA1. These 'clean' contacts then key your transmitter. D2 prevents inductive voltage spikes from damaging TR1, while C3 has a small additional integrating and filtering effect.

Fig 1. Basic layout. A keyed audio tone may be injected into the 'mic' or 'aux' socket of a cassette recorder

CONSTRUCTION
I used the 'ugly bug' wiring method to mount all components on a small piece of scrap PCB which was then fitted inside what was the battery compartment. [A stripboard layout – Fig 3 – is included for those who prefer this construction method. – Ed]

COMMISSIONING
Apply a 6V supply to the circuit and, using your key and an audio-frequency (AF) oscillator at around 1kHz, check that the relay

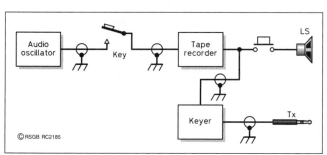

© RSGB RC2185

contacts are following your keying envelope. It might be useful to note the AF input level required to give satisfactory operation.

Then fit the PCB [or stripboard] and RLA1 inside the cassette recorder, ensuring that nothing can come into contact with any of the battery connectors. Leave a length of screened wire from D1 and ground, ready to make a connection to the recorder's loudspeaker output terminals.

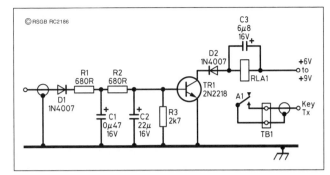

Fig 2. The auto keyer rectifies audio and uses it to drive a transistor which operates a relay

Having regard for the previous safety recommendation, three points must now be located on the machine.

1. The common negative ground connection.
2. The main positive low voltage supply rail (6 to 9V).
3. The 'live' side of the loudspeaker connection.

Points 1 and 2 should be readily found by locating the largest electrolytic capacitor, which is most likely to be the internal PSU reservoir.

Connect the keying circuit ground connection to the negative terminal on the capacitor, and the keyer's positive supply to the positive connection on the capacitor.

Connect the inner of the screened cable from D1 to the loudspeaker 'live' connection, the screen itself to ground.

Now double check for correct power supply polarity!

While making the loudspeaker connection, improvise a link or switch to silence the internal speaker when keying is in progress if this facility is not provided (some machines will do this if a dummy plug is inserted into an external LS socket).

Fig 3. Stripboard layout for the simple auto keyer. The underside of the board is not shown as no copper removal is required

Table 1. Components list	
Resistors	
R1, 2	680R, ¼W
R3	2k7, ¼W
Capacitors	
C1	0µ47
C2	22µ
C3	6µ8
All capacitors 16V tantalum bead if available	
Semiconductors	
TR	2N2218 (or similar)
D1, 2	IN4007 (or similar)
Additional items	
TB1	2-way terminal block
RLA1	5V model

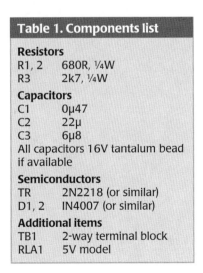

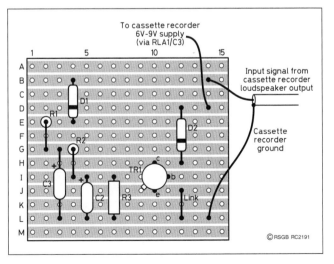

63

HINTS

A few practical hints may be useful at this stage:

- I have tried making a short recording on a one-minute endless loop tape, as used in telephone answering machines. These cassettes do not stand up to continuous running, so use any standard-length cassette.
- If an A-to-B dubbing machine is to hand, only a short length of error-free keying is necessary, since each repeated transfer from A to B can double the length of the recorded message.
- When making recordings, use a fairly high frequency (say 1kHz), and make connections via screened cables to avoid recording 50Hz onto the tape which would result in erratic keying. Remember that this simple circuit has no tone filtering. On replay, set the tone control for maximum treble to give some 50Hz rejection.
- If a microphone/mixer control is fitted, set this for minimum mix to avoid microphone sounds getting onto the tape.
- RF fields inside the shack are best avoided but we don't live in a perfect world. In worst-case conditions an unscreened/unfiltered recorder will rectify very strong RF voltages and gross keying instability will occur.
- Most recorders have an AGC system for controlling the record level so, assuming that the minimum input required is applied, there is no adjustment. On replay, the volume control must be set high enough to operate your keying relay. Set it to maximum if necessary, as the circuit is not likely to be overdriven.

Earth-continuity tester

When using mains-powered electrical equipment, a good-quality protective earth system is very important for safety. Good earth connections are additionally important for radio operation, both for protection against lightning strikes and also for the greater effectiveness of antennas that use earth as one half of a dipole.

In situations where the earth path is a *functional* earth as opposed to a *protective* earth, a simple low-voltage, low-current continuity tester or resistance meter is usually sufficient for checking earthing resistance, but for a proper test of a protective earth a high-current tester is needed. This is because a deteriorating earth connection in the form of a stranded wire where many of the strands are broken will still show a low resistance to a low-current tester but, in a fault situation when the earth path needs to pass a high current to ground and thus trigger a protective device, the high current causes the remaining strands to 'burn out', ie go open-circuit, before the protective device has time to operate; the protection is then non-existent.

SAFETY STANDARDS

Recognising this situation, the British and European safety standards for electrical safety, for example BS EN 60335-1 for household equipment, demand that the resistance of the protective earth path between an exposed metal part and the protective earth pin is less than 0.1Ω.

The equipment needed for checking to this standard is specialised and expensive, but this simple project provides a low-cost alternative and will check resistance at 2–3A if good-quality batteries are used. To simplify use, the circuit gives a pass/fail indication instead of a resistance value.

WHEATSTONE BRIDGE

The circuit can be considered in three parts; *test*, *detector* and *output indicator*. See the circuit diagram in Fig 1.

The *test* part of the circuit is based on a Wheatstone bridge, where the earth resistance path forms one of the resistance 'arms'. See Fig 2 for the principle behind a Wheatstone bridge. As a consequence of the values of resistance chosen (the test leads are assumed to have

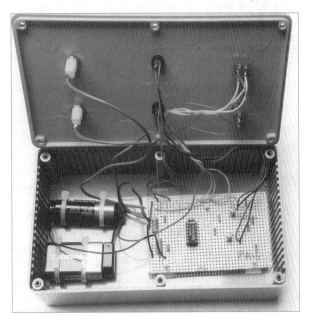

Inside a completed tester. Note that in a 'cased' project, the LEDs are removed from the stripboard and brought out to the front panel

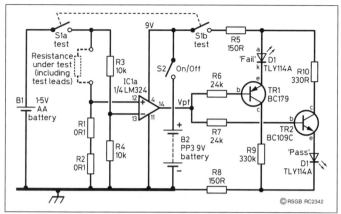

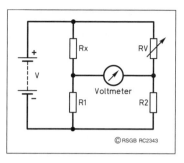

Above: Fig 1. Circuit diagram of the earth continuity tester

Above right: Fig 2. In a conventional Wheatstone bridge circuit, the value of an unknown resistance (Rx) is determined by adjusting a variable resistor (RV) with a calibrated scale until the reading on the voltmeter is zero. At that point (Rx/R1) = (RV/R2) so the value of Rx is then given by Rx=(R1×RV)/R2. The advantage of this method is that, at the balance point, no current flows through the voltmeter, so the resistance of it doesn't affect the measurement. This is a sensitive method for detecting small changes in resistance, as a small change causes a large meter reading

a resistance of 0.1Ω), if the earth resistance is less than 0.1Ω, the voltage between the midpoints of the two halves of the Wheatstone bridge will be positive, and if it is less than 0.1Ω, it will be negative. This is fed to the detector part of the circuit.

The *detector* is an op-amp wired as a comparator. Connected in this way, it has such a high gain that its output is roughly equal to either the positive or negative supply rail voltage, depending on whether the PD between its non-inverting and inverting inputs is positive or negative. It doesn't matter whether the PD is large or small – the output will always be at either extreme. This means there will always be a definite pass or fail indication from the detector, no matter how large or small the output from the Wheatstone bridge. This is important, as it means correct operation of the circuit doesn't depend on the voltage of the high-current battery, particularly as it is a chemical type whose output voltage can fall dramatically when a high current is being drawn. The pass/fail voltage V_{pf} from the detector then passes to the output indicator circuit.

The *output indicator* circuit consists of two LEDs, driven by transistors to provide sufficient current, which indicate either a pass or a fail for an earth path resistance of less than or more than 0.1Ω. TR2 is an npn transistor which switches on when its

Table 1. Components list
Resistors
R1, 2   0R1, 2.5W
R3, 4   10k
R6, 7   24k
R5, 8   150R
R9, 10   330R
All resistors metal oxide 0.4W 1%, except R1 & R2
Semiconductors
IC1   LM324
D1, 2   TLY114A yellow, or TLR114A red and TLG114A green
TR1   BC179 (general-purpose pnp)
TR2   BC109C (general-purpose npn)
Additional items
B1   1 × AA Duracell
B2   PP3
S1   Double pole, momentary on, or push-to-make
S2   SPST
Battery clips/holders
Stripboard
Plastic case*
2 × 4mm plugs & sockets*
2 × crocodile clips
* Only required if you are building the project in a case.

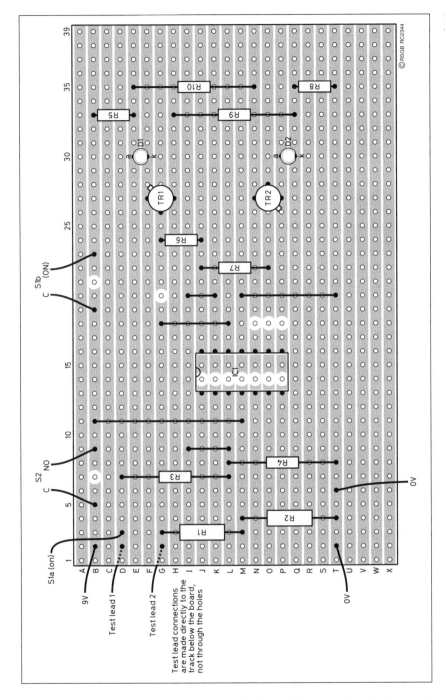

Fig 3. Layout of the circuit on stripboard

input is high, while TR1 is a pnp type which switches on when its input is low. A separate supply voltage is needed for the op-amp and LED circuit, since the test battery voltage will drop under a heavy load current.

Because the output of the op-amp does not swing completely to the

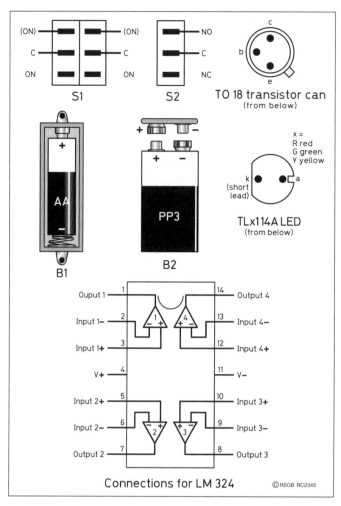

positive and negative supply rails, measures need to be taken to ensure that the LED driver transistors switch off correctly.

CONSTRUCTION

A suitable stripboard layout is shown in Fig 3 on the previous page, and Fig 4 shows how to identify and orientate several of the components. Use thick wire for the test leads!

HOW TO USE

Using flying leads with suitable connectors, eg crocodile clips, connect the circuit to each end of the earth path to be tested. This would usually be the mains plug earth pin and any metal part meant to be earthed. Then press the test button. Release the test switch as soon as a pass/fail indicator lights (certainly within 5 to 10 seconds, to lengthen battery life and prevent possible over-heating of R1 and R2).

SAFETY NOTICE

The project described here may be used to test the resistance of appliance earth connections,

Fig 4. Orientation (and pin-outs) of the batteries, IC1, LEDs and transistors

but it is *not* intended to conform to any *legal* requirements for the testing of electrical safety. The RSGB and the author accept no responsibility for any accident or injury caused by its use. *Never* use on mains equipment plugged into the mains – the connection to the mains plug earth pin mentioned in the previous paragraph implies that the plug is free. – *Ed*

An 80m transceiver

This transceiver was developed as a project that would hopefully appeal to the relatively inexperienced constructor, although it is probably not ideal for the complete beginner. The ability to handle a soldering iron and identify components is necessary, as is the ability to read and work from a circuit diagram, but a lot of detail is included to make construction as straightforward as possible.

DESIGN PHILOSOPHY

The aim was to produce a basic transceiver with minimum component count that required little in the way of tools or test gear to get working, but which would give creditable performance.

A tall order, maybe. The first step was to decide exactly what was required. The obvious choice was a single-band QRP rig. On the receiver side I opted for direct conversion, as a superhet design would be far more complex. For the same reason I decided on CW rather than SSB.

Now, if you are going to operate QRP CW for the first time, the ideal place to do it (in my opinion) is on 80m. On this band, a watt or two will give you QSOs all over the UK and well into Europe, a fact that I discovered around 30 years ago, before QRP operation really took off. I reduced the output of an old valve transmitter to about 3W, and to my surprise found that it seemed just as easy to make QSOs with this as it did with my 100W transmitter.

There are many circuits for this kind of transceiver around. A lot of them seem to be severely limited by over-simplification. I consider that VFO (variable frequency oscillator) control is highly desirable, but that RIT (receiver incremental tuning) is absolutely essential. If you are using crystal-control for transmit and receive, you have to rely on the other station being slightly off your frequency in order to produce the necessary beat note, or be able to shift the frequency of your oscillator slightly. With RIT and a VFO, it is a

The completed transceiver (centre)

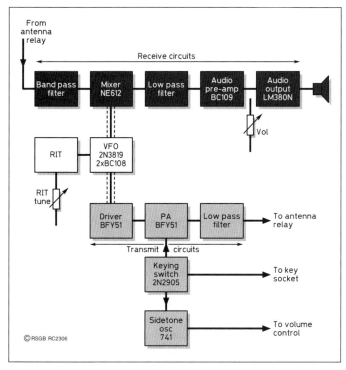

Fig 1. Block diagram of the 'Easi-Build' transceiver. Power supplies have been omitted for clarity

simple matter to 'net' onto another signal, and then adjust your RIT for a comfortable note.

Another highly desirable feature is a sidetone oscillator. This enables you to hear what you are sending, and does not add greatly to the overall complexity.

The method of construction is at least as important as the electronics in a practical project. A major difficulty to many constructors appears to be metalwork. 'Chassis bashing' went out with the widespread use of valves, and the skill has been largely lost. For this reason I opted to use double-sided copper-clad board to make a chassis/panel arrangement, with a rear apron and a screened enclosure for the VFO. Cutting and drilling are kept to a minimum, and the need to manufacture printed circuit boards avoided by mounting most components (with the exception of those in the VFO, which are mounted on a small piece of matrix board) in a 'dead bug' fashion. More on that later.

DESCRIPTION

The block diagram of the transceiver is shown in Fig 1. The VFO is common to transmit and receive.

On receive, the incoming signal is passed from the antenna socket by the changeover relay to the band-pass filter (which greatly attenuates out-of-band signals) and then to the mixer where it is mixed with the VFO signal to produce an audio beat note. Most operators favour a note of 800–1000Hz, so the VFO will be offset by this amount. Because there will be many in-band but unwanted signals at the mixer input, there will be many more at the output after the mixing process. Although most of these will be beyond the range of hearing, some may be strong enough to overload the audio stages. The low-pass filter will all but remove those at radio frequencies, and also attenuate the higher audio frequencies. The audio preamp now amplifies the wanted signal to a sufficient level to drive the output stage.

On transmit, the VFO signal is amplified by the driver and then the PA to give an output in excess of 1W. Harmonics are attenuated by the low-pass filter and the signal is passed to the antenna socket by a change-over relay. Keying is by means of a 'keying switch' (electronic, not mechanical) which keys the 13V supply to the PA. This keyed 13V also supplies the sidetone oscillator, the output of which passes to the audio output stage via the sidetone level control.

The RIT in this design allows the operator to tune approximately ±1.5kHz of the frequency on receive. The position of the RIT control does not of course affect the transmit frequency.

THE VFO

There is nothing remarkable about the VFO (Fig 2). Similar circuits are to be found in many pieces of equipment. The oscillator transistor is a 2N3819 field-effect transistor (FET).

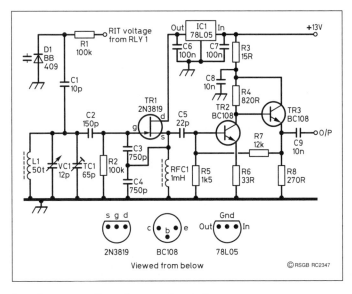

Fig 2. The VFO consists of a buffered Colpitts oscillator

This operates at a low power level so the generation of excess heat, the enemy of VFO stability, is minimised. The supply is stabilised at +5V by IC1. The buffer stage consists of two BC108s, chosen because they are cheap, common, and I have a bag full of them. Most amateurs must have a few lying around. The coil L1 is wound on a T50-2 toroidal core. Wound with the number of turns specified, you should have a VFO that works on the correct frequency, which might not be the case if any old former and core were used (I don't know of any current source of new coil formers). In practice, frequency stability is quite adequate.

Mention must be made of the tuning capacitor VC1. The tuning range is determined by the value of this component. 12pF allows coverage of the whole of the CW segment with a little to spare at each end, while 10pF will just cover the required range. As I prefer to have a little overlap, I used 12pF. With the amount of bandspread that this provides, the tuning rate is comfortable without a slow-motion drive. This, of course, simplifies mechanical construction. The value of a variable capacitor can be reduced by removing some of the plates. Actually, it is the number of gaps that determines the value, so if you divide this number into the value, you will know how many picofarads each gap contributes. The excess plates should be carefully removed with long-nosed pliers, taking care not to damage the remaining ones or over-stress the bearings or shaft. So, if you have a component with a value that is a little too high, you can still use it, although I wouldn't recommend trying to reduce the value by more than 50% as the results become less predictable.

It was decided *not* to build the VFO 'dead bug' fashion for two reasons: (a) the rigidity (very important in a VFO) wouldn't be as good as it could be; and (b) thermal stability. With the frequency-determining components in contact with the chassis, they would be more prone to sudden changes in temperature than if they were mounted independently. As printed circuits had been

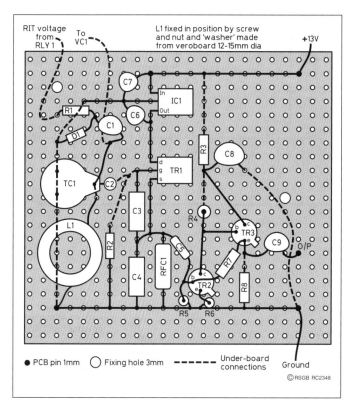

RIT voltage from RLY 1 | To VC1 | L1 fixed in position by screw and nut and 'washer' made from veroboard 12-15mm dia | +13V

● PCB pin 1mm ○ Fixing hole 3mm - - - - Under-board connections Ground

© RSGB RC2348

Fig 3. The VFO is built on matrix board

Fig 4. Circuit of the receiver. The VFO input is via a short length of miniature coax from C9

ruled out, the VFO was built on a 2¼in × 2½in piece of 0.1in matrix board. This, in turn, was mounted using two 12mm M2.5 screws, with extra nuts as spacers to keep the board clear of the chassis. See Fig 3 for the layout of the VFO board.

It turned out to be essential that the VFO be fully enclosed, not for the obvious reason of RF screening but because the slightest draught caused unacceptable drift. Holes are drilled in the screens to allow connections to pass through. Obviously this should be done before assembly. Initially the lid can be left off. Decoupling capacitor C40 should be mounted close to where the lead carrying the RIT voltage enters the enclosure.

THE RECEIVER

Referring to Fig 4 and the detail in Fig 5, the bandpass filter comprises T1, T2, C10, C11 and C12. The transformers are from Toko, which make circuits such as this easily reproducible.

Next comes the mixer. The NE612 is a useful device, containing an oscillator as well as a mixer. I decided not to make use of the oscillator in this case. The NE612 operates from an 8V supply provided by IC2, a 78L08.

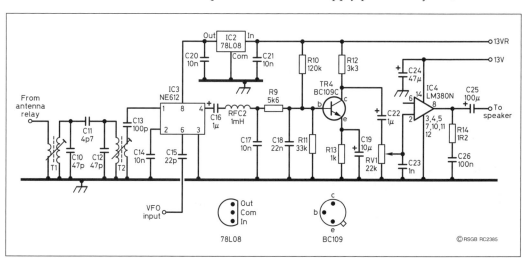

© RSGB RC2385

As mentioned previously, the output contains many signals, audio and RF. The RF signals are unwanted and are removed by RFC2 and C17. The higher audio frequencies are progressively attenuated by R9 and C18, which form a low-pass filter. This gives 3dB attenuation at 1.3kHz and 6dB per octave thereafter.

TR4 is the audio preamplifier. A BC109 was chosen rather than a BC108 because it gives more gain, which is required here.

The audio output stage is an LM380N. As shown in Fig 6, it is mounted upside down in the 'dead bug' style.

Unlike the previous stages, it is supplied with 13V on transmit as well as receive, as it is needed to amplify the sidetone signal to loudspeaker level.

Fig 5. T1 and T2 are mounted upside down on the earth plane and wired as shown

TRANSMITTER

Referring to Fig 7, when the key contacts are closed, RLY2 switches the antenna to the PA stage and 13V from the receiver to the transmit circuits. The RIT control on the front panel becomes no longer operative, and the VFO offset voltage now comes from RV3.

The VFO signal is fed to the input of the driver stage, a BFY51. The collector is coupled to the PA by L2, a toroidal transformer. It is more usual to tune this with a trimmer but (on 80m at least) the tuning is very flat, so the number of turns is optimised for a standard-value capacitor, in this case 150pF (C29). There is adequate drive over the whole of the tuning range.

Fig 6. IC4, the audio output chip, is mounted 'dead bug' style on the ground plane and wired as shown

Low-frequency parasitic oscillations, normally below 100kHz, sometimes occur in the driver stages of QRP transmitters, and can go unnoticed by the operator. I remember coming across a rough CW signal on about 3630kHz. I identified the station concerned, tuned down the band and also found him on 3560kHz. The two signals were 70kHz apart and, as expected, I found another rough CW signal

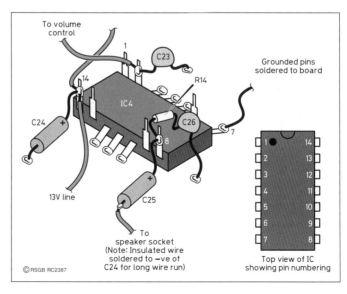

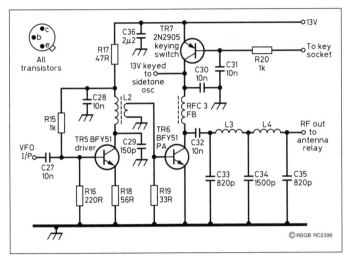

at 3490kHz. The station concerned was only running 3W and was nearly 200 miles away. This problem is normally caused by poor decoupling and is easily cured. I have had no such problems with this design.

The output from TR5 is taken via a link winding on L2 to the base of TR6, the PA. This is a BFY51 and produces in excess of 1W output. The DC feed choke RFC3 is not critical.

Some variation in wire thickness and number of turns shouldn't cause any problems,

Fig 7. The transmitter produces over 1W output and, thanks to adequate decoupling, is quite stable

so just try what you have available. Take care, though, not to damage the enamel insulation when passing the wire through the bead, as this is easy to do.

The output is coupled by C32 to the low-pass filter, comprising L3, L4, C33, C34 and C35. It then passes via the changeover relay RLY2 to the antenna socket.

TR7 is the keying switch. Its use is preferable to directly keying an RF stage such as the PA or driver, which can be unpredictable because RF has a tendency to find its way into keying lines. Also, there is the added advantage that the key or keyer only has to cope with the base current of TR7, a few milliamps.

The disadvantage of this kind of circuit is that it appears to be very good at injecting RF into supply lines and causing a shift in VFO frequency on 'key down' – something which is all too common in simple QRP equipment. The solution, again, is adequate decoupling, which is taken care of by C30 and C31.

Fig 8. How the RIT voltage to the VFO is switched between transmit and receive

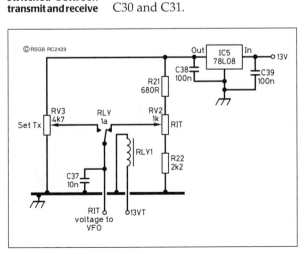

RIT

The requirement here is that the VFO frequency can be varied over a small range without moving the main tuning control. As you can see from Fig 8, RV2, the RIT control, is active on receive only. On switching to transmit, the RIT control has no effect and the frequency reverts to that set by the main tuning. The actual shift in frequency is accomplished by a BB409 varicap (variable capacitance) diode coupled to the VFO tuned circuit. When this is reverse-biased (ie positive to the cathode) it exhibits a capacitance which is dependent on the voltage so, by varying the voltage, it will tune the VFO over a small range. A fixed, stable voltage is required on transmit. This is provided by RV3. On receive, a variable voltage is provided by RV2, the front-panel-mounted RIT control. The supply to RV2 and RV3 is stabilised by IC5. Relay RLY1 selects the voltage from RV2 or RV3, depending on whether the transceiver is in the receive or transmit mode.

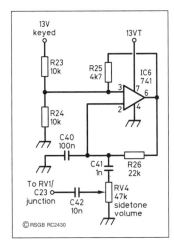

Fig 9. The sidetone oscillator and its associated level control

SIDETONE

The sidetone oscillator (Fig 9) uses a 741 IC. It draws its supply from the 13V line, but is only activated on 'key down' by 13V from the keying switch (TR7, shown last month). The output is taken to the volume control RV1 via the sidetone level control RV4 which is used to reduce the level of signal reaching the volume control. This is for operator comfort.

POWER SUPPLY

Power is supplied to the transceiver via a phono socket on the rear apron (see Fig 10). This is decoupled by C43. There is a diode, normally reverse biased, connected directly across the supply input. This is for protection, so if the supply is accidentally connected the wrong way round the diode will be forward biased and will conduct, blowing the 1A fuse, which should be installed in a holder in the power lead.

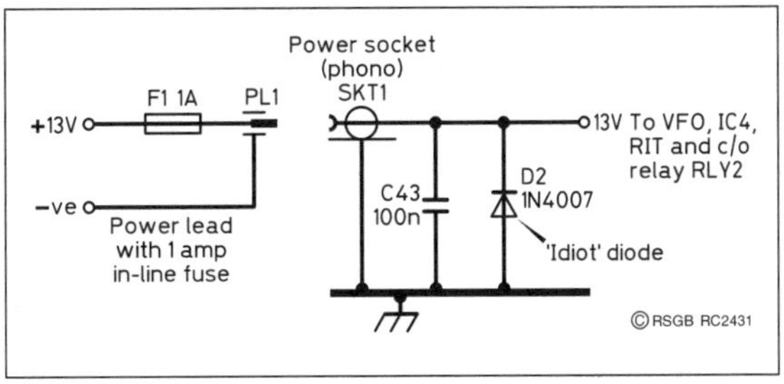

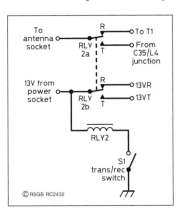

ANTENNA SWITCHING

This is accomplished using a double-pole changeover relay RLY2 (see Fig 11).

Fig 10. The connection to the supply contains protection against the wrong polarity power accidentally being applied[

MECHANICAL CONSTRUCTION

As stated previously, this project was designed for easy construction. Mechanical work has been minimised; nonetheless some is required. The copper-clad board should be double-sided glassfibre but SRBP could be used. This would be easier to cut but the end result is not as rigid. If you have, or if you know someone who has, access to a workshop guillotine, this part of the project could be very easy.

If not, you will have to cut the board by hand. Possibly the best way to do this is clamp it in a vice between two pieces of angle iron which are then used as a cutting guide.

Using a hacksaw, cut the pieces very

Fig 11. Transmit-receive switching

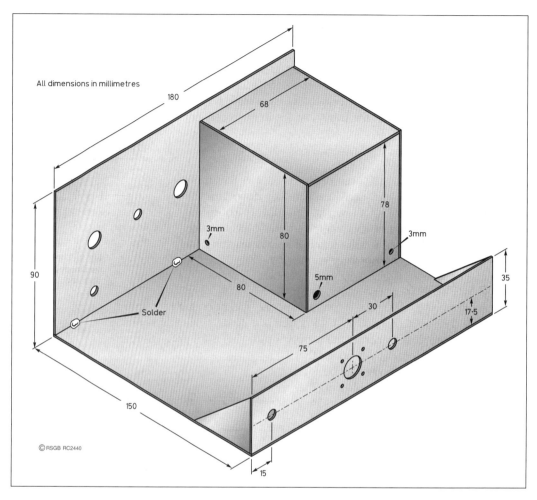

All dimensions in millimetres

180

68

3mm

3mm

78

80

5mm

90

Solder

80

35

30

17·5

75

150

© RSGB RC2440

15

Fig 12. The chassis, seen from the back

slightly oversize, then file down to the required dimensions. Refer to Fig 12 and Fig 13 for the dimensions.

When you have cut the base, front panel, rear apron, VFO enclosure and the two triangular supports, the holes can be drilled. First, mark their locations accurately and use a centre punch.

Start by drilling all holes 3mm in diameter. The larger holes can then be drilled to 6mm. At 10mm, more care is needed, as the drill can easily bite into the PCB. If you can, clamp the board firmly and drill through as slowly as possible.

The antenna socket can be any type you choose. If you want to fit an SO239, a 16mm hole will be needed. This can be made by marking out, drilling to 10mm and enlarging with a file. Alternatively, drill a series of small holes (say, 2mm) inside the circumference of a 16mm circle, cutting out the centre and finishing off with a file.

Finally, drill the two 3mm holes in the base for the screws which support the VFO board. In this instance, there is no need to measure – just place the matrix board that you will be using in position, and mark through the holes.

ASSEMBLY

With everything cut out and drilled, assembly can begin. Start by mounting the front panel on the base. Initially, use just one blob of solder in the centre. Then fix the side pieces of the VFO enclosure, again soldering lightly. After this, solder the rear apron in place. Finally, solder the two triangular supports in place. Do not mount the rear of the VFO enclosure at this stage. Inspect your work and if happy apply more solder to the joints and also solder in more places.

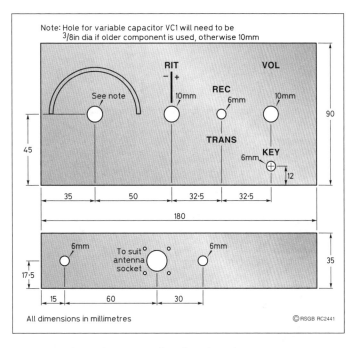

Note: Hole for variable capacitor VC1 will need to be 3/8in dia if older component is used, otherwise 10mm

All dimensions in millimetres © RSGB RC2441

Fig 13. Front panel and rear apron layouts

The end result will look more pleasing if the front panel is faced with card or thick paper, bearing labelling for the controls, along with a tuning scale for the VFO, and zero and ± marks for the RIT control. This can be done at any stage and fixed with adhesive.

ELECTRICAL CONSTRUCTION

This is best done in a logical order, rather than haphazardly. Refer to Fig 14, and start by mounting the large components: controls, sockets, switches, relays (the latter can be fixed in place with a spot of Super Glue®). Draw pencil marks on the base, along the line of the receiver (30mm from the side edge) and transmitter (30mm from the back edge). These are used as guides for the smaller components.

The first major job is the VFO. Fit all components and make all connections, taking great care to get the pin connections of the semiconductors correct. Check and double-check everything. Mount the VFO board, ensuring that the connections on the underside are clear of the base. Connect the variable capacitor using stiff wire (1.25 or 0.9mm). The lead to the RIT pin can be grounded for now. Connect a 13V supply and check that there is 5V at the output of IC1. All being well, the tuning range can now be roughly set. With VC1 at half-mesh, connect a short length of wire to the output (C9). Using a test receiver tuned to 3550kHz, adjust TC1 until a signal is heard. This is all that is needed at the moment. Don't be concerned if the VFO is not very stable at this stage.

Now for the receiver. Referring to Fig 4, start by assembling the bandpass filter (Fig 5) and fix it in place. Continue with IC3, IC2 and the smaller components. IC3 is mounted in the same manner as IC4 ('dead bug', ie legs

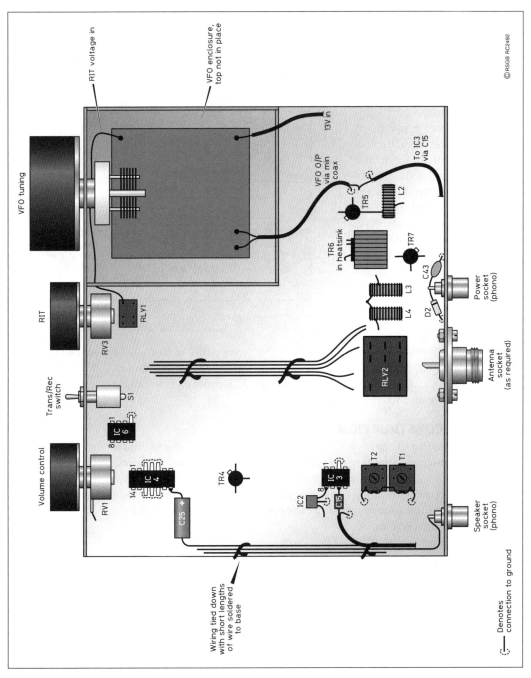

©RSGB RC2492

RIT voltage in

VFO enclosure, top not in place

13V in

VFO tuning

To IC3 via C15

VFO O/P via min coax

TR5

L2

TR6 in heatsink

TR7

C43

L3

L4

Power socket (phono)

RIT

RLY1

RV3

D2

Antenna socket (as required)

Trans/Rec switch

S1

RLY2

IC 6

Volume control

IC 4

TR4

IC 3

T2

T1

IC2

C15

RV1

C25 +

Speaker socket (phono)

Denotes connection to ground

Wiring tied down with short lengths of wire soldered to base

Fig 14. Layout of the major components. Smaller components are soldered between the larger ones

upwards), but only pin 3 is grounded. Where grounding is necessary, leads are soldered directly to the base. Carry on until the receiver is complete. The volume control is connected with miniature screened audio cable, and the output from IC4 is connected to the loudspeaker socket.

The receiver can now be tested using a temporary connection from the

VFO to IC3, with 13V to IC4, TR4 and IC2, and a length of wire as an antenna to the input of the bandpass filter. Sensitivity will be low, but it should be possible to hear something by tuning the VFO up and down its range.

Continue with the auxiliary circuits – the RIT, sidetone, changeover relay RLY2 [1] and the power connector. Permanent power connections can be made to all stages, a length of miniature coax connected to C15, and the wiring tied down neatly as shown. Now carry on and complete the transmitter and make a thorough check of everything.

TESTING

If at any stage a problem appears, turn off the power and investigate it. Connect a loudspeaker and an antenna, and turn on. Advancing the volume control should result in noise from the loudspeaker. Set the RIT control to zero. Tune the VFO and search for a steady signal at the centre of the tuning range. When one is found, using a suitable tool, adjust the cores of T1 and T2 for maximum volume. Now check the function of the RIT control.

Disconnect the antenna and connect a Morse key and a dummy load [2]. Set RV3 to the centre of its travel. Switch to transmit and press the key. The sidetone should be heard in the speaker. Holding the key down, tune the test receiver until the signal is found, somewhere near 3550kHz. If you have a power meter, it can be connected between the transceiver and dummy load. It should show an output of at least 1W.

SETTING UP

There are only three adjustments to be made – the VFO, RIT and bandpass filter – and they have already been roughly set.

VFO

Close the vanes of VC1 fully. The tuning knob should be set to exactly nine o'clock. Now set the test receiver to 3490kHz (if VC1

Table 1. RIT, sidetone, switching and power input circuits components list

Resistors

R21	680R
R22	2k2
R23, 24	10k
R25	4k7
R26	22k
RV2	1k
RV3	4k7
RV4	47k

Capacitors

C37, 42	10n ceramic disc
C38–40, 43	100n ceramic disc
C41	1n ceramic disc

Semiconductors

IC5	78L08
IC6	741
D2	1N4007

Miscellaneous items

RLY1	12V single-pole change over relay
RLY2	12V double-pole change over relay
F1	1A fuse with in-line holder
S1	SPST switch
PL1	Phono plug
SKT1	Phono socket

Table 2. Transmitter components list

Resistors

R15, 20	1k
R16	220R
R17	47R
R18	56R
R19	33R

All resistors 0.6W metal film

Capacitors

C27, 28, 30–32	10n
C29	150p
C33, 35	820p
C34	1500p
C36	2µ2

Inductors

L2	42 turns of 0.375mm enamelled wire on a T50-2 core, plus a 5-turn link of PVC-covered wire over the main winding
L3, L4	22 turns of 0.56mm wire on T50-2 core
RFC3	20 turns of 0.19mm wire on ferrite bead

Semiconductors

TR5, 6	BFY51

Table 3. Receiver components list

Resistors

R9	5k6
R10	120k
R11	33k
R12	3k3
R13	1k
R14	1R2
RV1	22k log pot

All resistors 0.6W metal film unless specified otherwise

Capacitors

C10, 12	47p ceramic plate
C11	4p7 ceramic plate
C13	100p ceramic plate
C14, 17	10n ceramic disc
C15	22p polystyrene
C16, 22	1μ, 16V electrolytic
C18	22n ceramic disc
C19	10μ, 16V electrolytic
C20, 21	10n ceramic disc
C23	1n ceramic disc
C24	47μ, 16V electrolytic
C25	100μ, 16V electrolytic
C26	100n ceramic disc

Inductors

RFC2	1mH
T1, 2	Toko KANK3333R

Semiconductors

TR4	BC109
IC2	78L08
IC3	NE612
IC4	LM380N

Table 4. VFO components list

Resistors

R1, 2	100k
R3	15R
R4	820R
R5	1k5
R6	33R
R7	12k
R8	270R

All resistors 0.6W metal film

Capacitors

C1	10p ceramic plate
C2	150p close tolerance polystyrene
C3, 4	750p close tolerance polystyrene
C5	22p polystyrene
C6, 7	100n ceramic disc
C8, 9	10n ceramic disc
VC1	10 or 12p (see text)
TC1	65p (Maplin WL72P)

Inductors

L1	50 turns of 32SWG on T50-2 toroidal core
RFC1	1mH

Semiconductors

TR1	2N3819
TR2, 3	BC108
IC1	78L05
D1	BB409

is a 12pF component) or 3500kHz if it is 10pF. Adjust TC1 until the signal is heard. Set the tuning control to the three o'clock position and check the frequency of the VFO by finding the signal with the test receiver. If using 10pF for VC1, the frequency should be about 3600kHz; if using 12pF it should be about 3610kHz. The VFO enclosure can now be completed but don't use too much solder on the lid – you might want to remove it at some future time. The VFO must be calibrated, but not yet. Since you have just heated everything up with a soldering iron, now is not a good time – allow several hours to elapse first.

RIT

Set the VFO to the centre of its range, the RIT control to zero (centre), and tune the VFO to give a beat note on the test receiver. Switch to transmit and adjust RV3 for exactly the same note. Switch between transmit and receive and carefully adjust RV3 so that the note doesn't change.

Bandpass filter

With the antenna connected, find a steady signal in the centre of the tuning range and carefully adjust T1 and T2 for maximum signal strength. This time take care and make sure it is right.

Calibration

It just remains to calibrate the VFO scale and the transceiver will be ready for use.

Before using it in earnest, though, it would be a good idea to get a local amateur to listen to your signal to make sure that all is well.

AND FINALLY . . .

The transceiver is delightfully easy to use. To net onto a station, start with the RIT at its centre position and adjust the main tuning for zero beat. The RIT can now be set to give the desired note.

With a reasonable antenna (eg a G5RV) you shouldn't have any shortage of QSOs. You are unlikely to achieve WAC or DXCC (please prove me wrong) – on the other hand, you will almost certainly have no problems with TVI or BCI. Have fun!

Rear view of the completed transceiver

NOTES

[1] As a particular relay is not specified for RLY2, you will have to work out the connections for yourself. If no data is available, this can be done visually and confirmed with a test meter on a resistance range.

[2] This can be three 150Ω 0.5W resistors in parallel.

A portable three-element 6m Yagi antenna

So, you've just acquired a transceiver with HF and 6m coverage, and you are keen to try out the 'magic band' for the first time. What better way than with a simple, portable antenna, which you can dismantle and use for some /P operation during the summer months?

REQUIREMENTS

When contemplating the addition of another antenna, most people would list their requirements as follows:

- Boom length, maximum 2m
- Reasonable gain
- Low cost
- Ability to dismantle easily for transportation
- Easy assembly with the minimum of tools
- Easy and quick tune-up
- Lightweight

First, materials are needed. A visit to the local D-I-Y shop should provide a good source with a wide range of 1m and 2m lengths of aluminium tubing and box sections. Use 12.5mm diameter tube in 2m lengths for the centre sections of the elements, and 10mm for the element end sections. These sections of tube fit snugly. You will also need a 1m length of 10mm rod to make the gamma match section. A 2m length of 25.4mm square U-section aluminium was used for the boom.

THE DESIGN

A three-element beam always has one reflector, one driven element and one director. For simplicity, 'plumbers delight' construction is employed so all the elements and the boom are at a common earth potential. This reduces some of the static electricity which can be prevalent in other types of design. However, this limits the feeding arrangements to basically a delta or a gamma match. The gamma match was chosen because of its easy adjustment, bearing in mind the requirements for quick tune-up and transportability.

DRILLING THE BOOM

The first thing to do is to measure out the boom and drill the holes for the reflector and director elements (see Fig 1 for dimensions). The 12.5mm holes are marked and drilled with a smaller (6mm) drill bit before the larger holes are drilled. Care must be taken to get these holes straight and level in both directions, otherwise the antenna will look very odd indeed. Using a small round file, file a notch in the top side of each of the four holes. The notch should be just large enough to pass the head of the screw which holds the

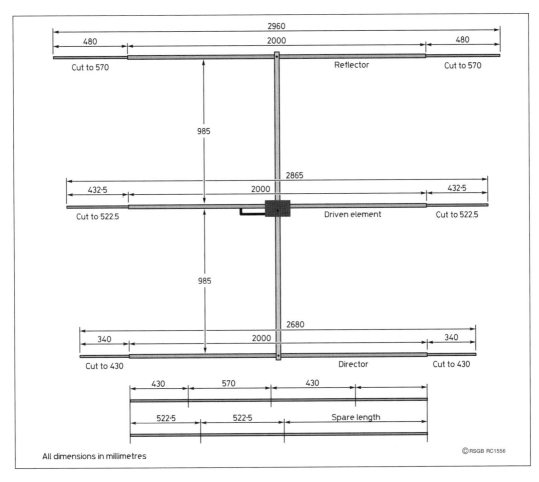

All dimensions in millimetres

10mm tubing to the 12.5mm tubing (see Fig 2). The reason for doing this is simple; it makes it easier to dismantle or assemble the elements on the boom. In the flat side of the boom drill a 2mm hole on the element centre line; this is for the screws to hold the elements in place (see Fig 2).

Fig 1. Dimensions and tube-cutting directions

DIRECTOR AND REFLECTOR

The 'standard' sizes of tubing available are perfect for a 6m beam. I bought two 2m lengths of 10mm tube and carefully cut them to the dimensions shown in Fig 1. In this way one piece gave me the correct lengths for four end-pieces (reflector and director). With these end-pieces cut to the right lengths, they were simply inserted into the 2m lengths of 12.5mm tubing and, after marking the correct length of the whole element, 2mm holes were drilled for the self-tapping screws to be driven through both the 10mm and 12.5mm tubing to hold them together. Each screw was driven through the tubing in a

Fig 2. Details of the element holes (reflector and director)

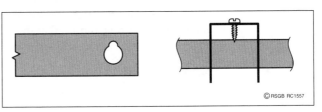

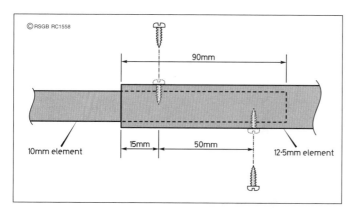

Fig 3. Detail of joining end-pieces to the main elements

similar manner on opposite sides. This is done on the director as well as the reflector (see Fig 3).

DRIVEN ELEMENT

The driven element is made of the same materials as the reflector and director. It is fixed to the top of the boom together with a plastic weatherproof box, which houses the 50pF capacitor and the gamma arm assembly.

The fixing screw through the box into the boom allows the driven element 12.5mm tubing to go in parallel to the boom for easy transportation.

At this stage, the 10mm tubing and the 12.5mm tubing are not yet fixed together. This is so that the 12.5 mm tubing will fit tightly into the connections box. Holes must be drilled in the 12.5 tubing of the driven element for the gamma arm. These holes allow connection of the gamma arm by a screw inside the tubing. A larger hole (8mm) is drilled through one wall of the tubing, and a 2mm hole is drilled in the opposite wall. The reason for the 8mm hole is to allow a screwdriver access to tighten the element-to-gamma-arm screw. After fitting and testing, the 8mm hole can be filled with putty or simply taped over with weatherproof tape. To get the correct location of the gamma arm holes, first find the exact centre of the driven element and then measure out 305mm – this is the centre-line for the holes. At the centre of the element, another 2mm hole must be drilled for the shield connection of the coaxial cable.

CONNECTIONS BOX

The connections box needs to be drilled to take the driven element, gamma matching arm and the capacitor shaft. The first holes to mark and drill are for the driven element. These were carefully marked on each end of the bottom half of the plastic box, then drilled with a 6mm pilot drill and then again with the 12.5mm bit. The gamma arm hole is drilled 40mm away from the element hole, and is only drilled in one end of the box. The driven element and 1m aluminium rod are placed into the box and a suitable position for the capacitor located and marked for drilling. The capacitor came from the junkbox, and the exact position is not critical as long as it does not foul the element or gamma arm (see Fig 4). A hole is required to screw the connections box to the boom; this is a 2mm hole drilled in the centre of the box. This screw goes through the plastic box and into the boom from the top side. The other hole required at this time is one for the coaxial cable. This is drilled in the same end as the gamma assembly hole in the centre of the box.

GAMMA ARM

With the 1m length of aluminium rod, measure and cut 320mm off. At one end, drill a 2mm hole for the connection point of the capacitor. 90° out from this, at the other end of the rod, drill a 2mm hole straight through the rod. Using a 6mm drill bit, drill half way through the rod to make a countersunk hole in the rod (see Fig 5). This is the attachment point for the gamma connecting arm.

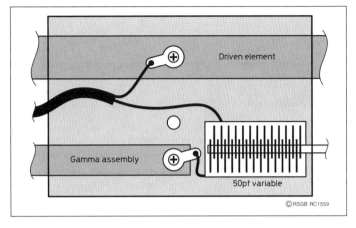

Fig 4. Connections from the coax to the gamma match

From the left-over length of aluminium rod cut a piece exactly 40mm long. With a file, one end must be filed to fit the gamma arm, and the other end filed to fit the 12.5mm tubing. If a drill-press is available, it can be used to make a perfect job by drilling through the rod with the correct size of drill bit. However, the file also works well, though it does take a bit longer. In each end of the gamma connecting arm a 2mm hole is required for the fixing screws (see Fig 5).

Using a self-tapping screw, screw the gamma connecting arm to the gamma arm, forming an 'L'-shaped gamma assembly. Place a screw through the 8mm hole in the driven element and screw the gamma assembly to the element using a self-tapping screw (Fig 6).

Slide the connections box over the driven element with the hole for the gamma assembly facing the gamma assembly. Insert the gamma assembly into the hole in the box and adjust the box until it is in the centre of the element. Now screw the end-pieces into the driven element in the same manner used for the director and reflector (see Figs 4 and 5). Fit the capacitor to the box and connect one side of the variable capacitor to the gamma assembly, using the 2mm hole (drilled earlier) and a solder tag held in place with a self-tapping screw. Feed the coaxial cable into the box from the outside and strip the ends ready for connection. The shield of the coaxial cable is soldered to a solder tag and screwed to the driven element using a self-tapping screw. The inner of the coaxial cable is soldered directly to the other

Fig 5. The gamma arm assembly

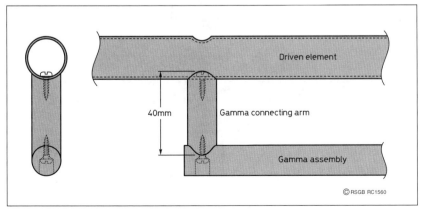

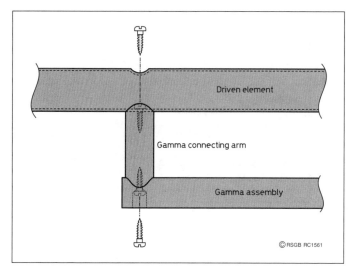

Fig 6. Connecting the gamma arm assembly to the element

side of the variable capacitor. All the external connections to the box can now be weatherproofed using hot-melt glue, epoxy or similar product.

FIXING THE DRIVEN ELEMENT

Locate the centre of the boom and drill a 2mm hole in the exact centre of the flat edge (see Fig 7). The connections box is screwed to this point using a self-tapping screw. Now, turn over the boom and align the driven element so that the boom and the element are at 90° to each other. Drill through the boom and the plastic box and into the driven element with a 2mm drill bit. This hole will be off-centre and it holds the driven element in the correct place when the beam is being used.

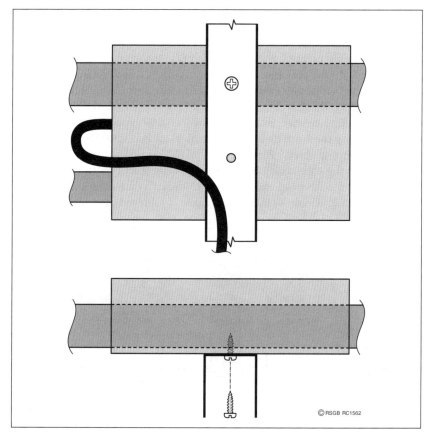

Fig 7. Connecting the box to the boom

WEATHERPROOFING

Fill the ends of all the elements with hot-melt glue, epoxy, or a similar long-lasting product to keep rain out of the tubing. After checking all the connections are good in the connections box, close the lid and seal against the weather. If required, drill holes in the side of the boom for a mast clamp.

Table 1. Components list	
Boom section	One 2m length, 25.4 × 25.5mm square 'U' aluminium
Element centres	Three 2m lengths 12.5mm OD aluminium tubing
Element ends	Two 2m lengths 10mm OD aluminium tubing
Gamma match	One 1m length 10mm aluminium rod
Connection box	Plastic box 70 × 122 × 50mm
VC1	Variable capacitor, 50pF
Other items	Solder tags, self-tapping screws

CHECKING THE SWR

The SWR should be quite good if the dimensions in this article are followed closely. Place the assembled antenna in a clear area at least 3m off the ground and check the SWR on a known SWR bridge. The SWR can be adjusted by turning the capacitor and checking again. It is best to check the SWR at both band edges and set the SWR minimum at the centre of the band (see Fig 8 for SWR readings on the three beams built for testing purposes).

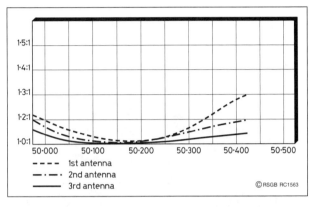

Fig 8. SWR readings on the three beams built for testing purposes

ASSEMBLING AND DISMANTLING

Lay out all the metal pieces in a clear flat area. Rotate the driven element to 90° and insert the retaining screw through the boom and into the driven element. Slide the director through the boom and lock it in place using a screw through the boom and into the element. Do the same thing with the reflector. The complete antenna is shown in Fig 9.

To dismantle the beam, lower the mast, remove the antenna and the three screws in the boom. Remove the director and reflector and rotate the driven element through 90°. Only a single screwdriver is required for the beam and three screws.

Fig 9. The complete three-element 50MHz beam

CONCLUSIONS

The three-element beam meets all the criteria which were set out at the beginning. It has proved robust enough for everyday use at home as well as portable use on my 10m pump-up mast. The total weight for the antenna is less than 3kg! The total cost of the beam was £18.63.

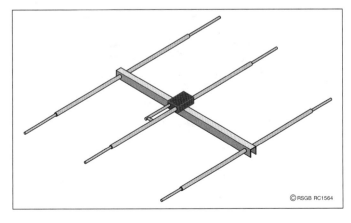

Transmit-receive changeover

When you build a transmitter, you need a method of switching the antenna from the receiver to the transmitter during the transmit period. It is also useful to co-ordinate the transmitter power supply switching with the antenna changeover switching. The transmitter should not have its power connected during receive periods; if you were to touch the key by accident, the transmitter will operate without any load. In many cases this will cause the PA transistor to fail and may harm your receiver.

DOUBLE-POLE SWITCH

The simplest changeover is a double-pole switch. This can change the antenna with one pole and the power with the other. The disadvantage, however, is that you must remember to flick the switch each time you want to change from receive to transmit.

With a small QRP transmitter it is possible to flick the little rig right off the bench!

A primitive method of sharing the antenna between the transmitter and the receiver is to have the antenna permanently connected to the transmitter. The receiver is then connected to the antenna via a small capacitor. Two back-to-back diodes across the receiver input are used to prevent large RF voltages from damaging the receiver. The diodes can be 1N4148 or similar and the capacitor can be 50 to 100pF. The problems with this arrangement are that some transmitter power is dissipated in the diodes, and a lot of RF still gets into the receiver front-end; on transmit loud thumps are heard from the receiver. The volume control on the receiver can be turned down but this can become a tiresome chore when using this method.

ELECTRONIC ANTENNA CHANGEOVER

A much better method is shown in Fig 1. The circuit causes a relay to change over when the key is pressed. The circuit is simple and uses a pnp transistor as a switch and an npn transistor as

The completed unit

Fig 1. Circuit diagram of the electronic transmit/receive antenna changeover switch

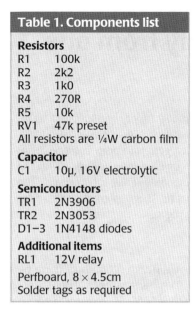

Table 1. Components list

Resistors
R1   100k
R2   2k2
R3   1k0
R4   270R
R5   10k
RV1   47k preset
All resistors are ¼W carbon film

Capacitor
C1   10μ, 16V electrolytic

Semiconductors
TR1   2N3906
TR2   2N3053
D1–3 1N4148 diodes

Additional items
RL1   12V relay

Perfboard, 8 × 4.5cm
Solder tags as required

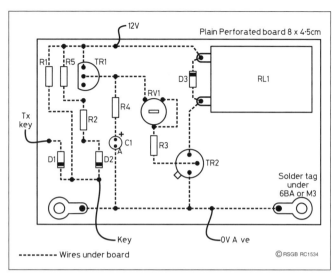

Fig 2. Component layout

a relay driver. The relay can be any small one with a 12V coil. Take care to check that the relay does not already have a diode across the coil. If it does you can leave it in and omit D3, but take care to fit the relay the correct way round or it will not work.

The circuit is built on plain perforated board, as shown in Fig 2, with the components pushed through and solder connections made beneath the board. The solder tags act as earth contacts to brass standoffs to the case when the board is bolted to them. The changeover board can be housed in the same case as the transmitter.

When complete, check the wiring and connect the 12V supply. The relay should change over when the key is pressed. The delay is set by RV1. The relay should drop out when you pause between words but not between letters.

Diode D1 is included to prevent interaction with the circuits to be keyed on the transmitters.

Dual-voltage supply from one battery

Many projects require a positive and a negative supply voltage, which can of course be provided by using two batteries or a dual power supply. A neater (and cheaper) method, however, is to generate both voltages from a single battery. One way of doing this is to use a simple voltage divider of two resistors but unfortunately this method is only really satisfactory if the circuit doesn't consume any significant current. This design allows current to be drawn without significantly affecting the supply voltage.

THE POTENTIAL DIVIDER

Fig 1(a) illustrates the principle of the potential divider. On the left are two 5kΩ resistors connected in series across a 10V power supply. There are no prizes for saying that the voltage at their junction is 5V. If we were to ground this middle point, we would have a dual power supply of ±5V. *This only works, however, if neither side of the 10V supply is grounded.*

To call this a dual power supply is a little misleading. Even though a multimeter connected across each resistor would read 5V, we cannot draw any power from the circuit without drastically affecting its output voltages.

Here is an example. In Fig 1(a), a load of 500Ω is placed across one half of the supply. The parallel combination of the 5000Ω and 500Ω resistors has an effective value of about 455Ω. This can now be considered to be in series with the other 5000Ω resistor across the 10V supply. This now draws, not 1mA as originally, but 1.8mA. (You can verify these figures for yourselves.) We now need to ask ourselves what is the voltage across the 500Ω load resistor? Well, the current is 1.8mA and the resistance is 500Ω, so the voltage is just over 0.9V. So you can see that this simple system cannot deliver any power into an external circuit. Because the original power supply is still delivering 10V, you can see that the negative supply (after connection of our load to the positive half) will now be 10V – 0.9V = 9.1V. Not only has the positive voltage virtually collapsed but the negative voltage has shot up to almost 10V! This system will not fulfil our purposes.

The situation can be improved somewhat if the potential divider is made from lower resistances, as in Fig 1(b). You may

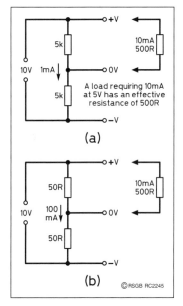

(a)

(b)

©RSGB RC2245

Fig 1. The effect of a load on a voltage divider. (a) A voltage divider of 5k resistors adequately provides V+ and V– of ± 5V when there is no load, but if an effective load of 500Ω is connected across the upper resistor it effectively becomes 5kΩ in parallel with 500Ω, ie around 450Ω, and V+ drops to less than 0.5V. (b) If 10 times the required load current flows through the potential divider, ie the load presents 500Ω in parallel with 50Ω, the upper resistor is equivalent to about 45Ω and the loading effect is not so great. Even so, V+ will drop to around 4.5V

like to work out the corresponding voltages here, with and without the load. You will see that the voltage drop on connection of the load is smaller, but look at the current that flows down the divider chain – 100mA – even when there is no load!

This is a very inefficient system which wastes battery energy and of course shortens its life, and also needs resistors capable of dissipating the significant heat generated by the wasted power. If the circuit presents any significant current drain, for example to drive a LED or an audio amplifier, the supply voltage will drop significantly in one or both supply rails.

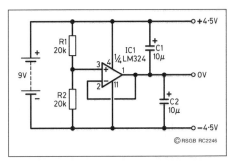

Fig 2. Circuit of the dual-voltage supply

AN ELEGANT SOLUTION

An effective solution to this problem is to use the potential divider to generate a reference voltage at half the battery supply voltage, but to incorporate another circuit element that prevents the divider supplying any power. This extra element is an operational amplifier (op-amp) connected as a voltage follower, to provide the required load current at this voltage. In this way, the power for the load comes directly from the main power source through the op-amp, not from the potential divider.

As before, we can use the output of the op-amp as our zero voltage level, and the two connections from the original power supply now become the positive and negative rails. This is shown in Fig 2. A battery is ideal as the power source because, as we noted above, neither side of it must be grounded if a true dual-voltage supply is required.

CIRCUIT DESCRIPTION

An op-amp connected this way is using 100% negative feedback, which means that all of the output voltage is fed back to the inverting input. This gives a voltage gain of unity, and means that the output voltage is always the same as the input voltage (hence the name *'voltage follower'*).

However, although it is called a 'voltage follower' it is actually a current-controlled device. A change in voltage at the op-amp output caused by a change in load current also appears at its input due to the feedback. This causes the op-amp to change its output current in such a way as to counteract the voltage change on the output, and keep it fixed at the reference voltage. This change occurs effectively instantaneously, and means the circuit responds to all changes in load current demand while at the same time keeping the supply voltages fixed.

VALUES

The component values are not critical, and simple rules-of-thumb are sufficient for choosing values. The main thing to consider is the load current required, which must be within the specification of the op-amp. For example, a typical use of this circuit would be for powering an LM324

Table 1. Components list	
Resistors	
R1, 2	20k, 0.4W, 1%
Capacitors	
C1, 2	10μ (see text) electrolytic, voltage rating to exceed the supply voltage
Semiconductor	
IC1	LM324 quad op-amp
Additional items	
Stripboard	
Battery clip	

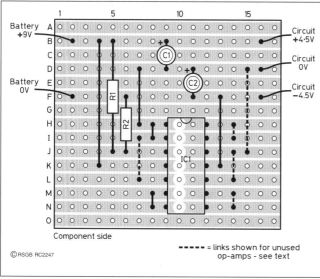

Fig 3. How to lay out the dual-voltage supply on stripboard

quad op-amp. This device will comfortably supply up to 20mA, which should be sufficient for most low-power applications.

The voltage divider resistors should carry around 10 times the input current for the op-amp, which is in the order of microamps, so any value between 10kΩ and 100kΩ will be fine. They do not need a greater power rating than that usually found in low-power circuits, so a 0.4W rating is more than sufficient. The accuracy of the resistor values determines how closely the values of the plus and minus supply rails match, ie how close the reference voltage is to exactly half of the battery voltage. 1% tolerance is certainly adequate.

The capacitors are smoothing capacitors, so again values are not critical, and could be increased to 100µF or even 1000µF without much increase in the physical size of the components. The voltage rating for the capacitors should naturally be higher than the output voltage.

STRIPBOARD LAYOUT

The circuit could be built as a 'stand-alone' project if required and used with different projects as needed. The component count is so low, however, that it's probably just as easy to build the power supply as part of a larger project.

Fig 4. Orientation and internal layout of the LM324 quad op-amp integrated circuit

The stripboard layout is shown in Fig 3 for an LM324 quad op-amp, one quarter of which generates the plus, zero and minus supply rails which run along the full length of the stripboard for easy linking to other components.

The other three are available for use in the circuit of which it is part but, if these are not used, it is good practice to tie the output to the inverting input, and the non-inverting input to 0V. This reduces the possibility of any unforeseen behaviour of the op-amps such as latching or self-oscillation, which may increase current consumption or have other unpredictable and undesirable effects.

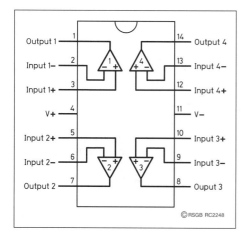

CONSTRUCTION

Construction is straightforward, the only things to watch out for being the correct orientation of the integrated circuit (check for the position of pin 1 – see Fig 4), and the correct polarity of the electrolytic capacitors (see the markings on their casings).

The J-pole antenna

The J-pole antenna is essentially an end-fed half-wave verti-cal dipole. Normally a half-wave antenna is fed in the centre where the impedance is low and it can be fed directly with coaxial cable. This method of feeding poses a mechanical prob-lem with a VHF vertical antenna and it is more convenient to feed it at the end. Because the end of a half-wave antenna has a high impedance, a transformer must be used to allow the antenna to be matched to low-impedance coaxial cable.

The half-wave antenna in Fig 1 is shown connected to a length of twin-wire transmission line which provides the nec-essary impedance transformation from the coaxial cable feed to the high-impedance end of the half-wave element. The bottom end of the transmission line transformer is at zero ohms so the coaxial cable has to be tapped some way up the trans-mission line at the 50Ω point.

The J-pole antenna was originally designed so that the 'RF earth' point was at earth potential. This can be difficult with some arrangements so a single radial, or counterpoise, is in-cluded to provide this RF earth.

CONSTRUCTION

Making the antenna is very simple. Take a 2m length of 300Ω ribbon feeder and measure 430mm from one end, cut one of the wires and then remove that section by cutting the webs. Care-fully remove 5mm of insulation from the uncut wire and a similar length from the cut end and join them together to make the connection marked 'RF earth' in Fig 1. From this point measure a further 430mm up the cable, cut one side and remove the top cut side of the ribbon. Finally, from the top cut end measure 960mm and trim off the end.

The feed is accomplished by connecting the braid of the coaxial feeder with the shortest possible tail to the RF earth point and the inner to a tapping point 90mm from the RF earth up the transformer section on its long side. These lengths were found by experiment and are a little shorter than expected. However, the VSWR is 1.4 at the centre of the band, rising each side, and so little improvement is expected by further optimisation.

My antenna is supported inside a 2m length of plastic water pipe with the coax hanging down by the counterpoise. The bottom is left open but the top is closed with a plastic cap; this way condensation can easily dry out. It is clamped to the supporting pipe using the last 150mm.

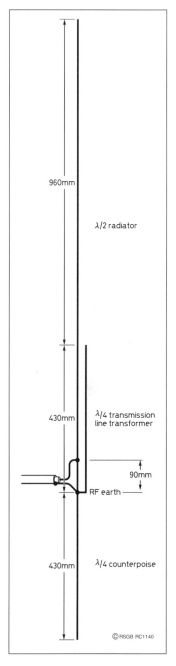

960mm

λ/2 radiator

430mm — λ/4 transmission line transformer

90mm

RF earth

430mm — λ/4 counterpoise

©RSGB RC1140

Fig 1. The J-pole antenna with di-mensions shown for 145MHz

Feedline verticals for 2m and 6m

During decades of portable operation on the 2m and 6m bands I have met amateurs who needed antennas that were efficient, simple to construct and also easy to hang up almost anywhere. The feedline vertical antennas described here provide an excellent match to a transceiver without a separate antenna tuner. They can be easily coiled-up and stored in your luggage or even stuck in your pocket, as they are made from only one piece of flexible coax cable. The basic design is extendable to any frequency segment between 50 and 150MHz.

The feedline vertical can be suspended by nylon line from a tree limb or any other convenient support. It should dramatically extend your maximum communication range when operating portable, when compared to a whip or 'rubber duck' antenna

DESCRIPTION

The basis of the design is the coaxial sleeve antenna (Fig 1), which was popular until the advent of modern SWR analysers. A true resonance will always be found, but I have never achieved a better SWR than 2:1 in such antennas, probably due to stray capacitance. The lesson learned, however, was that the RF current had no trouble in travelling up the inside of the coax and then making a 180° turn to travel back down the outer sleeve.

Because of this, perhaps we don't need the sleeve. Why not just use the braid of the coax itself? If we do this, however, how do we let the RF 'know' when it should stop flowing and reflect back towards the centre of the dipole, as it did when it came to end of the braid in the coaxial sleeve antenna?

After trying different wide-band devices, I found that a coaxial cable choke resonating within the band segment in question was the best solution to meet my requirement, and it is easy to work out reliable dimensions from the formulas in Fig 2.

2m ANTENNA

In the photos you can see the 144MHz version of this antenna. It is made from a 387cm-long (152¼in) piece of RG58CU coaxial cable, of which a quarter-wavelength (use the formula) of sheath and braid is stripped off, this forming the upper part of the dipole. Next, measure the lower part of the antenna (use the formula) and mark the starting-point of the choke. For the choke, wind 4.6 turns of the coaxial cable onto a piece of 32mm (1¼in) diameter PVC tube. The caps on each end are not essential but they are useful to centre the cable and lock the turns.

A ring terminal or tag needs to be soldered to the tip of the dipole, bearing in mind that this will lower the resonant frequency a little. Trimming, if necessary, should be done at the tip, outdoors, well away from objects that might affect the resonance. Don't cut more than 6mm (¼in) at a time. The

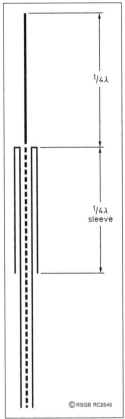

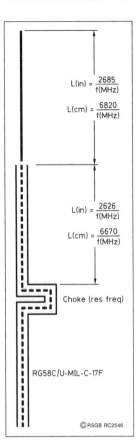

In Fig 2:

$$L(in) = \frac{2685}{f(MHz)}$$

$$L(cm) = \frac{6820}{f(MHz)}$$

$$L(in) = \frac{2626}{f(MHz)}$$

$$L(cm) = \frac{6670}{f(MHz)}$$

Choke (res freq)

RG58C/U-MIL-C-17F

¹/₄λ

¹/₄λ sleeve

SWR should be less than 1.3:1 and the impedance very close to 50Ω across the entire band. Observant readers will see that I have used exactly five electrical half-waves (340cm) of feeder. It is a good idea to make any additional feeder a multiple of 68cm (26½in).

6m ANTENNA

For a 50MHz vertical dipole, you should start with a 728cm (286½in) length of RG58CU. Using the formulae, follow the same constructional procedure as before. In this instance, the choke consists of 11.8 turns on 50mm (2in) diameter PVC tube.

Although not critical, you can centre the antenna to your favourite 6m frequency by cutting the tip little by little, and still enjoy a 1.3:1 SWR across the band. Adding feeder to this antenna should be in multiples of 198cm (78in).

Left: **Rolled up and ready to go –** the 2m version of the feedline vertical

Centre: **Fig 1. The** coaxial sleeve antenna. **The best SWR that could be obtained was 2:1**

Right: **Fig 2. The** feedline vertical. **The best SWR that could be obtained was 1:1**

Dry battery tester

The output voltage of a dry battery with no load, as measured by a digital voltmeter for example, will give an indication of the state of the battery. However, that is only part of the story – even a spent battery can show quite a high voltage when out of circuit but, as soon as an attempt is made to draw current, the voltage will fall dramatically.

You can check the current capability of a dry battery by connecting it directly across an ammeter but, as this is effectively short-circuiting it, it is not to be recommended. It not only wastes the energy of a good battery but can also cause overheating and other permanent damage if the short-circuit lasts for more than a second or two. A better method is to connect the battery to a load similar to the one it would experience in normal use and measure the current the battery is capable of providing. The device described here uses this method, and will test the current capability of the two most commonly-used dry cells, the 1.5V and 9V types, and provide a visual indication of the state of the battery via a series of LEDs.

HOW IT WORKS

For the test, the battery is connected to a load resistor capable of withstanding the current the battery can provide without 'burning out'. The voltage dropped across the resistor, which is proportional to the current (by Ohm's law), is compared with a series of reference voltages chosen to represent the changeover points between the current ranges of interest. Where the generated voltage exceeds the particular reference voltage, one or more of a series of LEDs light to give an indication of how much current the battery is supplying and hence whether it is in an 'excellent', 'good', 'adequate' or 'needs replacing' state.

THE LOAD

AA- and PP3-type Duracell batteries will provide up to 2A for intermittent use. A suitable test load that draws sufficient current to test the battery's state without wasting too much of its power will be one that allows around 1A to flow.

THE CIRCUIT

This is shown in Fig 1. R1 and R2 are the load resistors for 1.5V and 9V batteries respectively, and switch S2(a) selects whichever test is required. The same value of current passing through a 1Ω and a 4.7Ω resistor will generate a different potential difference (PD) across each resistor of course, and so for the indicator part of the circuit to be common to both tests, the PD across the 4.7Ω resistor is split by a potential divider consisting of R3 and R4. Using this method, the value of V_{test} is nominally 1V/amp of current supplied by the battery under test for both types of battery. S2(b) selects between the non-divided and divided source voltage as appropriate.

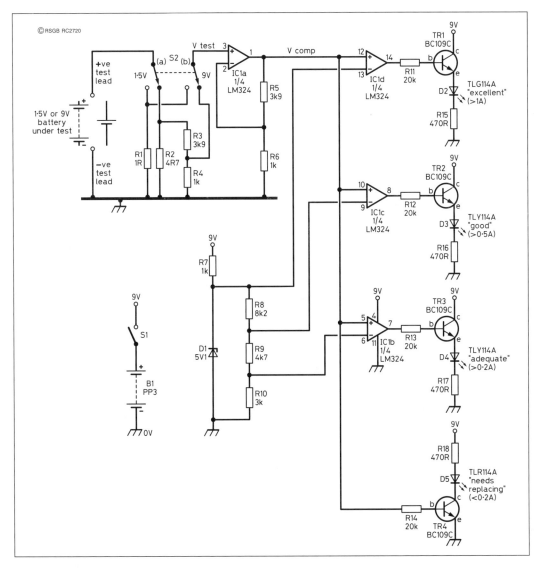

© RSGB RC2720

Before passing to the comparator stage, V_{test} is amplified by IC1a, connected as a non-inverting amplifier with a nominal gain of five. This means that the voltage applied to the comparators (V_{comp}) will range from 0V to around 6V, and give more predictable switching of the comparators by avoiding using voltages around or below the turn-on voltages of the IC's internal p-n junctions (about 0.7V).

The reference voltages against which V_{comp} is compared are generated by a voltage divider network consisting of R8, R9 and R10 across a 5.1V Zener diode D1. The use of a Zener diode ensures that the reference voltages remain stable, even if the power supply battery voltage changes with use. R7 limits the current through D1 to its working value. Reference voltages of 1V, 2.5V and 5.1V are used to give changeover points at nominally 0.2A, 0.5A and 1A.

Fig 1. The unit works by checking the voltage that a battery is able to deliver when a substantial load is placed on it

97

Fig 2. Stripboard lay-out of the dry battery tester

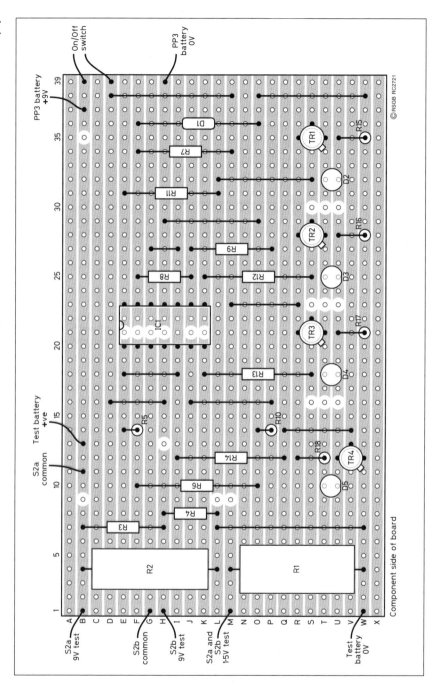

Indication of current values of less than 0.2A (but greater than approximately 0.14A) are given by the LED driven by TR4. This part of the circuit takes further advantage of the turn-on voltage of a p-n junction, in that TR4 will not switch on until V_{comp} is above around 0.7V. Hence there will be no false indication of a current of less than 0.2A when no battery is being tested,

Table 1. Components list	
Resistors	
R1	1R, 5W, see text
R2	4R7, see text
R3, 5	3k9
R4, 6, 7	1k
R8	8k2
R9	4k7
R10	3k
R11–14	20k
R15–18	470R
All resistors metal film, 0.6W, 1%, unless specified otherwise	
Semiconductors	
IC1	LM324
TR1–4	BC109C
D1	BZY5V1 0.5W Zener
D2	TLG114A green LED
D3, 4	TLY114A yellow LED
D5	TLR114A red LED
Additional items	
S1	SPST
S2	DPDT
Stripboard	
PP3 battery and clip	
Battery holders for test batteries, if required, see text	
Case to suit	

even if V_{comp} is not exactly zero due to any small offset or bias currents associated with the op-amps.

IC1b, IC1c and IC1d are op-amps used to act as comparators by not employing any feedback. Due to the high gain of op-amps used in this way, their output will be around 0V when the non-inverting (+) input is less than the reference voltage on the inverting (–) input. In this state, the transistor connected to its output (one of TR1 to TR3) will be 'off' and the associated LED (one of D2 to D4) will not light. As soon as V_{comp} rises above the relevant reference voltage however, the op-amp output will immediately go to around 9V, switching 'on' the appropriate transistor and lighting the associated LED. R11 to R18 limit the current to the working values of the transistors and LEDs.

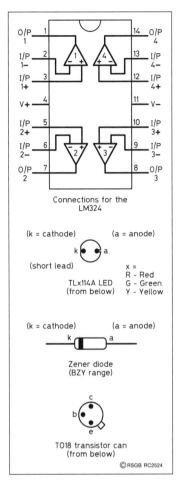

Fig 3. Orientation of the semiconductors

CONSTRUCTION

A suggested stripboard layout is shown in Fig 2. Construction is straightforward, the only particular point to mention being the correct orientation of the IC, transistors, Zener and LEDs. The markings that indicate their correct orientation are in Fig 3.

The components are all general-purpose low-power types, apart from the load resistors which ideally need to dissipate 5W (although 3W types should suffice as the current to be carried only passes for short periods).

It is useful to employ different colours for the positive and negative test leads as the correct test battery polarity must be observed for the LEDs to light correctly. Alternatively, solder battery cell holders and clips permanently to the ends of the test leads (if this is done only one test battery should be connected to any holder or clip at a time, and it should be removed immediately after testing).

BATTERY TESTING

With S2, select the 1.5V or 9V battery test as appropriate, switch on, connect the battery to the test leads and note which LEDs light.

Finding noise sources in the shack

The radio amateur of yesteryear is to be envied to some degree because of the relatively small amount of electrical noise that caused problems. How different it is today, with every house full of electrical equipment, all capable of emitting electromagnetic radiation to interfere with the poor radio amateur who is trying to listen to signals on the bands.

This project detects the radiation that causes problems to the amateur and the noise can be heard. When we say to other members of the household "Please don't turn on that computer, vacuum cleaner or TV" they cannot understand why we are complaining, but this little device will allow you to show them and let them hear the 'noise' with which we have to contend.

CONSTRUCTION

The circuit (Fig 1) uses a telephone pick-up coil as a detector, the output of which is fed into an LM741 IC preamplifier, followed by an LM386 IC power amplifier.

The project is built on perforated board (Fig 2), with the component leads pushed through the holes and joined with hook-up wire underneath. There is a wire running around the perimeter of the board to form an earth bus.

Build from the loudspeaker backwards to RV1, apply power and touch the wiper of RV1. If everything is OK you should hear a loud buzz from the speaker. Too much gain may cause a feedback howl, in which case you will need to adjust RV1 to reduce the gain. Complete the rest of the wiring and test with a finger on the input, which should produce a click and a buzz. The pick-up coil comes with a lead and 3.5mm jack, so you will need a suitable socket.

Fig 1. The detector works by receiving stray radiation on a telephone pick-up coil and amplifying it to loudspeaker level

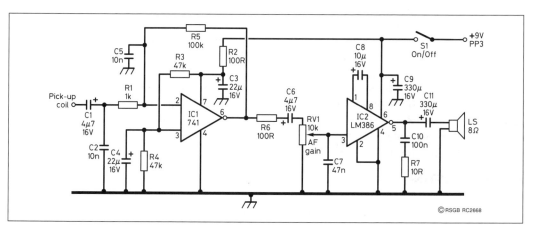

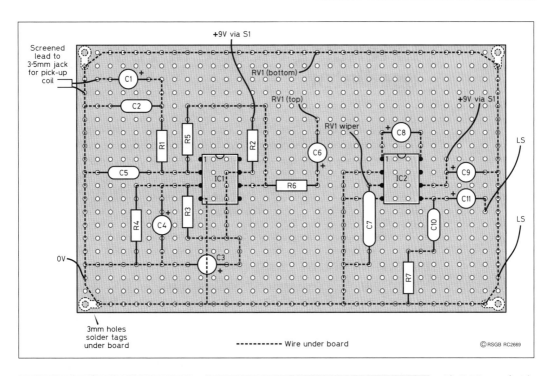

+9V via S1

Screened
lead to
3·5mm jack
for pick-up
coil

RV1 (bottom)

RV1 (top)

+9V via S1

RV1 wiper

C1 +

C2

C8 +

C6 +

C9 +

LS

C11 +

LS

C5

R6

IC1

IC2

C3 +

0V

3mm holes
solder tags
under board

------------ Wire under board

© RSGB RC2669

Fig 2. The project is built on perforated board with point-to-point wiring underneath

Table 1. Components list

Resistors

R1	1k
R2, 6	100R
R3, 4	47k
R5	100k
R7	10R
RV1	10k, with switch

All resistors 0.25W

Capacitors

C1, 6	4µ7, 16V electrolytic
C2, 5	10n
C3, 4	22µ, 16V electrolytic
C7	47n
C8	10µ, 16V electrolytic
C9, 11	330µ, 16V electrolytic
C10	100n

Semiconductors

IC1	LM741
IC2	LM386

Additional items

LS1   small 8Ω loudspeaker

Perforated board, 7 × 7cm
PP3 battery and clip
3.5mm mono jack socket
Case to suit
Telephone pick-up coil (Maplin)

Table 2. Readings obtained by placing the pick-up coil next to various household items

29MHz oscilloscope	0.56V
Old computer monitor	0.86V
Old computer with plastic case	1.53V
New computer monitor	0.45V
New tower PC with metal case	0.15V
Old TV	1.2V
New TV	0.4V
Plastic-cased hairdryer	4.6V
Vacuum cleaner	3.6V
Drill	4.9V

RELATIVE NOISES

I placed a high-impedance meter set to a low AC voltage range across the speaker leads, to give a comparative readout between different items of equipment in the home. The readings that I obtained are shown in Table 2.

Extending the use of your dip oscillator

The grid dip oscillator (GDO) offers a quick and easy means of checking (to a degree of accuracy acceptable for experimental purposes) the induct-ance value of coils in the microhenry (µH) range and capacitors in the picofarad (pF) range, such as are commonly used in radio circuits. This can be very useful, for example, when constructing an ATU, a crystal set, a short-wave receiver, a VFO or a band-pass filter for a direct-conversion receiver.

DETERMINATION OF L AND/OR C

For this purpose, I keep with my GDO two fixed-value RF coils of known inductance – 4.7µH and 10µH – and one capacitor each of 47pF and 100pF (but the choice of values is yours). You may decide to keep one or more of each, to be selected from Table 1 and Table 2.

My personal choice of coil type is the moulded RF choke (Maplin) or RF inductor (Mainline or RS). These are axial-lead, ferrite based, encapsulated, easy to handle, and readily available at low cost in a range of fixed values. The capacitors are 5% tolerance polystyrene, also axial-lead.

To determine or verify the value of either an RF coil or a capacitor, simply connect the unknown component in parallel with the appropriate known component to form a parallel LC tuned circuit, ie an unknown L in parallel with a known C (or vice versa), then use the GDO to determine the resonant frequency of the parallel LC circuit.

The value of the unknown component can then be obtained easily to an acceptable approximation, by using the relevant formula from Tables 1 and 2 and a pocket calculator.

Note that, in Tables 1 and 2, *F is the frequency in megahertz as given by the GDO.*

Example 1

An unknown capacitor in parallel with my known 10µH inductor produces a dip at 6.1MHz, hence $F = 6.1$MHz. From Table 1, the value of the unknown capacitor is given by:

$$C_{pF} = 2533 \div F^2$$
$$= 2533 \div 6.1 \div 6.1$$
$$= 68pF$$

Example 2

An unknown coil in parallel with my known 47pF capacitor produces a dip at 12.8MHz, hence $F = 12.8$MHz. From Table 2, the value of the unknown inductance is given by:

Table 1. To determine an unknown capacitance						
For known L (µH) of	1	2.2	4.7	6.8	10	22
Unknown C (pF)	$25330/F^2$	$11513/F^2$	$5389/F^2$	$3725/F^2$	$2533/F^2$	$1151/F^2$

Table 2. To determine an unknown inductance						
For known C (pF) of	10	22	33	47	68	100
Unknown L (µH)	$2533/F^2$	$1151/F^2$	$768/F^2$	$539/F^2$	$373/F^2$	$253/F^2$

$$L_{\mu H} = 539 \div F^2$$
$$= 539 \div 12.8 \div 12.8$$
$$= 3.3\mu H$$

Bear in mind that, because the accuracy of results relies upon the frequency as derived from the GDO, it would be sensible to keep the coupling between the LC circuit and the GDO as loose as possible, consistent with an observable dip. This minimises pulling of the GDO frequency. Also, rather than relying upon the frequency calibration of the GDO itself, it might be useful to monitor the GDO frequency on an HF receiver or a digital frequency meter.

A final point worth considering is that each fixed-value inductor of the type mentioned might have its own self-resonant frequency, but these would typically lie above the HF range so should not be a problem.

For example, the self-resonance of my own 10µH inductor is about 50MHz and that of the 4.7µH one is about 70MHz. You could quickly and simply find out the self-resonant frequency of an inductor by taping it to each of the GDO coils in turn and tuning across the full frequency span.

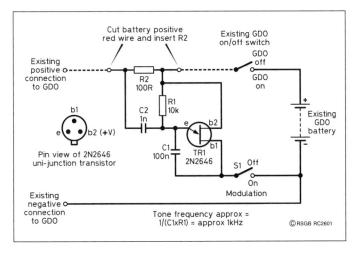

Fig 1. This audio oscillator adds a 1kHz AM tone to a GDO

It is best to make L and C measurements at frequencies much lower than the self-resonant frequency of your chosen test-inductor, but perhaps better be safe than sorry and stick with the lower microhenry values if your interest lies between 1.8 and 30MHz.

TONE MODULATION

Sometimes it is useful to be able to hear an audio tone when using the GDO as an RF signal source in association with a radio receiver.

If your GDO does not have tone modulation, you might like to construct the simple add-on 1kHz audio oscillator circuit

Table 3. Components list
Resistors
R1   10k
R2   100R
Capacitors
C1   100n
C2   1n
Semiconductor
TR1   2N2646

Fig 2. Stripboard layout

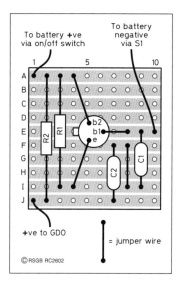

shown in Fig 1 on the previous page. It uses a unijunction transistor, the frequency of oscillation being given approximately by $1/(R_1 \times C_1)$. The 1kHz tone output connects via C2 to the positive supply line of the GDO, which it modulates.

R2 acts as the modulator load and its value helps to determine the level of modulation. This produces simple but effective tone modulation of the GDO's RF signal which can be heard on an AM or an FM receiver.

Fig 2 gives a suggested layout of the components on a small piece of stripboard, without the need for any track cutting. The finished board might conveniently mount on one of the GDO meter terminals, provided care is taken to isolate the copper-tracks from the terminal.

An op-amp tester

When building a circuit, it is not unusual to find that it doesn't work first time. After fault finding it's not unusual either to find that one of the supply rails had been connected to either an input or output of an operational amplifier (op-amp). This could be either because of an incorrect link, an unnoticed short-circuit between copper tracks, or a direct connection between the IC pins due to a 'whisker' of solder. With a stripboard project it could also be due to an intended break in one strip that has not been completely cut through, or a burr of copper that is shorting to an adjacent track. It might also be due to a required break that has been omitted.

When the fault has been rectified and the circuit still doesn't work, it then isn't possible to say whether this is now due to another fault present or whether it is because the op-amp was damaged by the initial fault. The best solution to this dilemma is to check independently that the op-amp is working correctly. This simple project will perform just such a check. It can also be used to check that an op-amp salvaged from unwanted equipment for use in another project works correctly. It will prevent time being wasted in checking for construction faults, when it is in fact the component that is faulty.

HOW IT WORKS

When a component fails, particularly a semiconductor device, it usually fails catastrophically, ie it fails completely and doesn't just work half-heartedly. In the case of an op-amp, this usually means that the output goes to the value of one supply rail or other and stays there. Another possibility is that the output goes to some fixed DC value between the two supply rail limits.

This circuit works by incorporating the op-amp under test into an astable

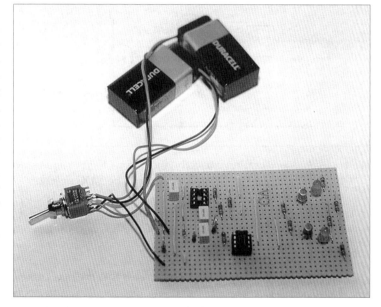

The completed op-amp tester

oscillator circuit. If the circuit oscillates, the op-amp is fine; if not, it is damaged. In order that an oscilloscope is not needed to observe the output waveform, the op-amp output is connected to a detector circuit and the output from this is connected to an indicator section. The indicator section drives

105

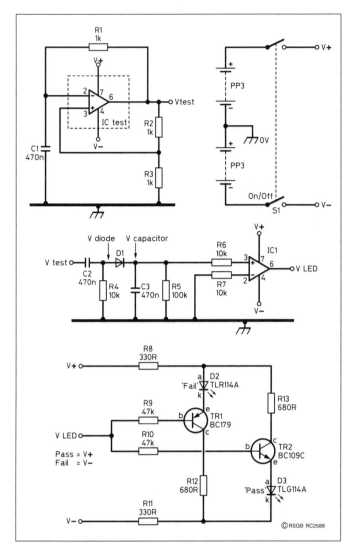

Table 1. Components list	
Resistors	
R1–3	1k
R4, 6, 7	10k
R5	100k
R8, 11	330R
R9, 10	47k
R12, 13	680R
All resistors are metal film, 0.6W 1%	
Capacitors	
C1–3	470n polyester film
Semiconductors	
IC1	LM741CN
TR1	BC179
TR2	BC109C
D1	1N4148
D2	TLR114A red LED
D3	TLG114A green LED
Additional items	
S1	DPDT
Stripboard	
PP3 battery and clip both × 2	

two LEDs: a green one to indicate that oscillation is present, ie the op-amp passes the test; and a red one to indicate no oscillation, ie the op-amp fails the test.

THE CIRCUIT

Components R1 to R3 and C1, in combination with the op-amp under test, form an astable oscillator (see Fig 1). When this is operating correctly, the output V_{test} is a square wave with a frequency of about 1kHz. Otherwise V_{test} will be a DC voltage.

Fig 1. Circuit diagram of the op-amp tester

If V_{test} is a DC voltage it will be blocked by C2 and V_{diode} will fall to 0V as the right-hand plate of C2 discharges via R4 (with time constant C2 × R4). D1 then isolates IC1 from this part of the circuit and both inputs of IC1 are connected via resistors to 0V. The non-inverting ('+') input, however, is connected via 110kΩ (R5 + R6), whereas the inverting ('−') input is connected via 10kΩ (R7). This means that the bias current entering the '−' input will be greater than that entering the '+' input, deliberately creating a differential input offset voltage. Because of this, and the high gain of an op-amp connected without any negative feedback resistors, the output of IC1, V_{LED}, will very nearly swing to the negative supply voltage. This switches TR1 on and TR2 off, and so the 'Fail' LED D2 lights.

106

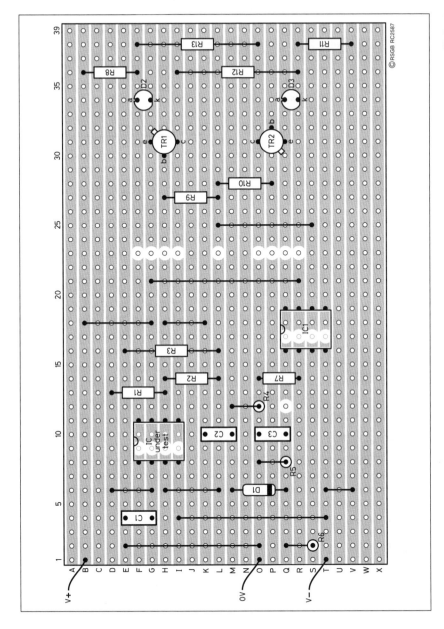

Fig 2. Stripboard layout and wiring diagram

If, however, V_{test} is a 1kHz square wave, C2 will not block the signal and D1 will conduct during the positive part of the waveform. This means C3 will charge up, as R5 is now acting as the discharge path for C3 and the time constant C3 × R5 is long compared with the period of the 1kHz signal. The resulting positive value of $V_{capacitor}$ causes the current flowing into the '+' input of IC1 to exceed the current flowing into the '−' input and V_{LED} swings nearly to the positive supply voltage. This switches TR1 off and TR2 on, and so the 'Pass' LED D1 lights.

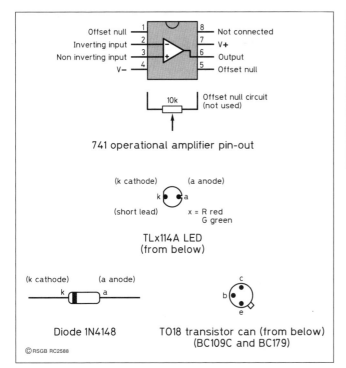

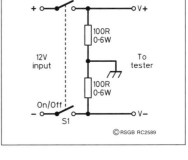

Above: **Fig 3. Orientation of critical components**

Right: **Fig 4. To power the op-amp tester from a 12V supply, substitute the top right-hand part of Fig 1 with this circuit. If you do this, remember that the 'chassis' of the tester (denoted by the chassis symbols) will not be at ground potential, so you will need to keep the unit isolated from ground**

CONSTRUCTION

The stripboard layout for the op-amp tester is shown in Fig 2. An 8-pin DIL socket is of course needed for the op-amp under test but IC1 can be inserted into a socket or soldered directly into place as preferred. It is always a good idea to install ICs in sockets though; as well as eliminating the possibility of damage due to soldering it makes removal for testing much simpler. Care needs to be taken to ensure that the op-amp, transistors, diode and LEDs are connected the right way round, and Fig 3 shows how to determine the correct orientations.

COMPONENTS

All the semiconductors used in this project are general-purpose types, and the values of the components associated with them are chosen to limit voltages and currents to their working values. Otherwise, values of capacitance and resistance are chosen to give suitable time constants to ensure reliable operation.

IN USE

Any single op-amp package with standard DIL pin-out connections that operates with a power supply of ±9V can be tested, which covers most situations. Simply insert the op-amp to be tested into the test socket, again ensuring correct orientation, switch on, and note which LED lights.

USING A 12V SUPPLY

If the use of two 9V batteries to power the op-amp tester doesn't appeal to you, it is possible to adapt it to run from a 12V power supply. To do this, simply take two 100Ω resistors, make a potential divider (as shown in Fig 4), and use this instead of the top right-hand part of Fig 1.

A colourful voltage monitor

Probably the most common unit used in electrical, electronic and radio engineering is the volt, which is a measure of the amount of energy associated with the electrical charge of electrons at a particular point. One volt is defined as one joule of energy per coulomb of charge, and there are two main situations where it is encountered. The first is where a voltage is 'dropped', for example across a resistor, and this is a measure of the energy lost as a current flows (ie electrons move) against that resistance. This is usually referred to as *potential difference* (PD). The other is where a device 'pushes' the electrons, in the form of an electrical current, around a circuit, as does a signal generator or a battery. In this case the voltage is usually referred to as the *electro-motive force* (EMF). So when measuring the voltage of a battery we are measuring how much energy the battery is capable of giving to the circuit of which it is part. As the energy stored in the battery is used up, its output voltage falls, an indication of the falling amount of energy available to 'drive' the electrons. Often the degree to which a battery has been 'used up' is of more interest than the actual voltage output, and in this case an indication of a voltage 'range' is more useful.

WHAT IT DOES

This project uses a series of LEDs to indicate when a voltage falls below 12V, when it is inside an acceptable range of 12 to 13V, when it is inside a second acceptable range of 13 to 14V, when it is a little too high at between 14 and 15V, and to warn when it is above a maximum acceptable 15V. With these ranges, it is ideal for monitoring car battery voltage, and when using recharge-

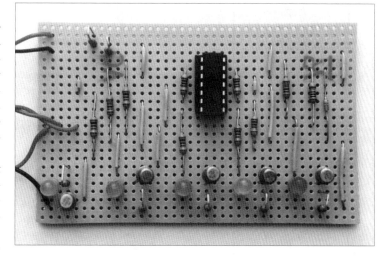

able batteries to power equipment around the shack or 'in the field'.

The completed unit. The components are laid out on stripboard. The monitor could be put into a project box or the LEDs mounted separately if desired

HOW IT WORKS

To give a meaningful output in the form of a sequence of ranges, a series of reference points is needed with which to compare the test voltage. These reference voltages can easily be generated using a resistance divider chain. However, the main reference from which all these are derived needs to be stable at all times, and it is not sufficient simply to connect the resistor chain

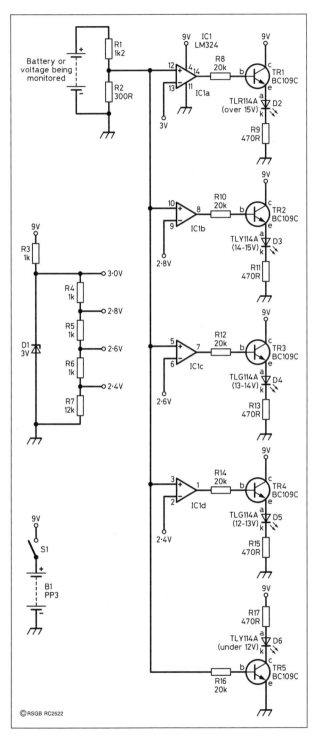

Table 1. Components list	
Resistors	
R1	1k2
R2	300R
R3–6	1k
R7	12k
R8, 10, 12, 14, 16	20k
R9, 11, 13, 15, 17	470R
All resistors metal film, 0.6W 1%	
Semiconductors	
IC1	LM324
TR1–5	BC109C
D1	BZYC3 Zener, 3V 0.5W
D2	TLR114A red LED
D3, 6	TLY114A yellow LED
D4, 5	TLG114A green LED
Additional items	
S1	SPDT
Stripboard	
PP3 battery and clip	
Crocodile clips	

Fig 1. The device works by comparing a divided fixed voltage to a divided sample of the voltage being monitored

across the supply battery. If this technique were to be used the reference voltage would of course change as the supply battery aged or when it was subject to different load currents. In this project a Zener diode is used to achieve a stable reference voltage. These devices are semiconductor diodes connected 'backwards', ie reverse biased, and when used like this the normal flow of current through them is blocked. However, all semiconductor devices have a 'leakage' current which is a small current that passes in the 'wrong' direction, and Zener diodes are designed to give a particular fixed voltage across them when this leakage current is flowing.

Each reference voltage from the resistance divider chain is fed to the inverting input of its own comparator, the test voltage being connected to all of the non-inverting inputs. Thus the output of an individual comparator

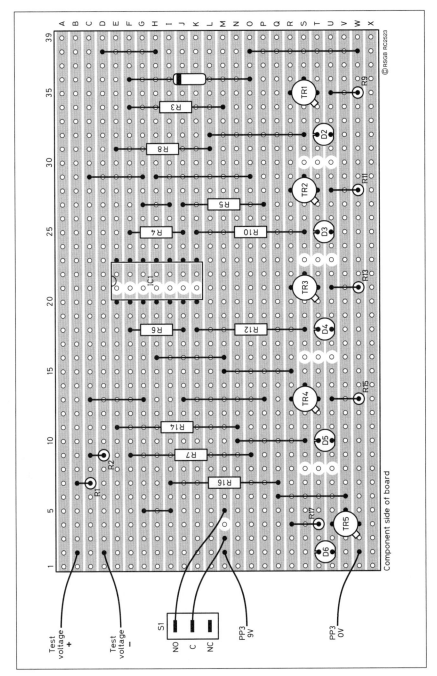

© RSGB RC2523

Fig 2. Stripboard layout of the voltage monitor

will be around 0V unless the test voltage rises above the comparator's reference voltage, at which point it will switch over to a value close to the positive supply voltage. By connecting each output to an LED with its own driver transistor to provide enough current, the LEDs will light up in turn as the test voltage increases.

111

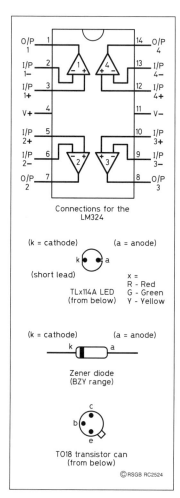

Connections for the
LM324

(k = cathode)   (a = anode)

k ● ● a

(short lead)   x =
                     R - Red
TLx114A LED   G - Green
(from below)   Y - Yellow

(k = cathode)   (a = anode)

k a

Zener diode
(BZY range)

c
b
e

TO18 transistor can
(from below)

© RSGB RC2524

**Fig 3. Semiconduc-
tor connections**

Since the test voltage will normally be higher than the circuit supply voltage, the test voltage and the reference voltages are both scaled down by the same amount, in this case by a factor of five. The test voltage is scaled by a simple voltage divider, the reference voltages by using a 3V Zener diode (the maximum reference voltage before scaling being 15V).

The '<12V' indication is given by switching the LED driver directly from the scaled test voltage. In this way a false '<12V' indication is not given when there is no test voltage connected.

CIRCUIT

The circuit diagram for the voltage monitor is shown in Fig 1 on p110. The comparator functions are obtained by using the operational amplifiers (op-amps) of IC1 wired without any feedback. In this configuration the op-amp's high gain means the output can only ever be equal to the positive or negative (0V) supply voltage, depending upon whether the voltage at the non-inverting input is more positive or more negative than the voltage at the inverting input.

The transistors all operate as switches, ie they are either 'on' (conducting) or 'off' (non-conducting). When they are 'on', the transistor passes enough current to light the LED (the op-amp output alone cannot reliably provide enough current to do this). The values of resistors chosen in the transistor/LED part of the circuit are those that limit voltages and current flows to the working values of the transistors and LEDs.

CONSTRUCTION

The stripboard layout for the project is shown in Fig 2 on the previous page. The LM324 quad op-amp integrated circuit, the transistors and the LEDs all need to be connected the 'right way round', and Fig 3 shows the correct orientation of these devices. An important point to note is that the 'positive' test lead needs to be connected to the more positive terminal of the battery under test, so it is useful to use red and black wires respectively for the test leads.

COMPONENTS

An LM324 quad op-amp was chosen as it gives four devices in a single package and can be powered from a single supply voltage. The transistors are a general-purpose npn type, and similarly the LEDs are different colours of a general-purpose type. The lower the tolerance value of the resistors used for the voltage divider chain, the more accurate are the reference voltages.

IN USE

Simply connect the test leads to the voltage being monitored (ensuring that the 'positive' test lead is connected to the more positive terminal), and switch on!

A direct-reading capacitance meter

Bags of surplus unused capacitors from rallies become an attractive proposition provided you can fathom out what they are. You may also possess polystyrene capacitors where the marking ink (used to show the values and working voltage) has rubbed off. This instrument saves you trying to recall myriads of capacitor value codings. You will find other uses, for example the ability to check the setting obtained with a variable capacitor used to find a resonance. The reading obtained by this direct-reading capacitance meter gives the value needed for a substitute fixed capacitor.

THE CHOICE

There have been various approaches made in designing capacitance checkers, including:

Multi-range test meter

A suitably-scaled multimeter can give approximate values for capacitors, using what is really one of the AC voltage test ranges in conjunction with a mains-powered step-down transformer. Such a solution is messy but reasonably effective. It is not a recommended practice with low-voltage capacitors or for use by beginners!

Test meter with internal ac source

Many modern digital multimeters have capacitance ranges. An alternating voltage is generated by an oscillator powered from internal batteries. These meters have their uses, though, but are expensive. However, the display format provided, which usually consists of three-and-a-half digits, is unsuitable for taking variable readings, as when checking a suspect plastic dielectric tuning capacitor.

Service technician's resistance/capacitance test bridge

Resistance/capacitance ratio bridges used to be one of the tools of every service technician. These versatile units often have additional features. There is usually a source of high DC voltage (0 to 500V variable) for the forming of high-voltage electrolytic capacitors. This voltage may also used for testing leaking paper capacitors which, as every service technician knows, contribute to the demise of resistors followed by valves and transformers.

Purpose-designed capacitance checker

A direct-reading capacitance meter has the facility to transfer capacitive reactance, in linear fashion, to the scale of an analogue direct-reading meter. The analogue meter is equally happy with fixed or variable capacitance readings.

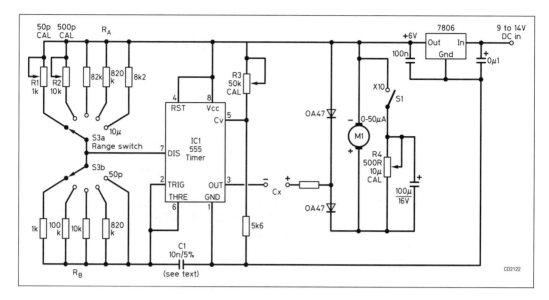

Fig 1. Circuit diagram of the capacitance meter. The resistor groups R_b and R_b are discussed in the text, as are the pair of resistors selected by switch S3. S3a selects R_a and R_b is selected by S3b. All fixed resistors are 5% ¼W carbon types, and polarised capacitors are electrolytics

The design of cheap and accurate direct-reading capacitance meters becomes a lot easier with the availability of solid-state integrated circuits. The simplest designs settled on the use of the 555 timer IC operated as a monostable oscillator.

The definitive work was done by A Wilcox for the magazine *Television* in the early 'seventies. Constructional projects subsequently appeared in *Electronics Australia* October 1976, and in *QST* January 1983. The design used for this project begins with the *QST* circuit and then sets out to overcome many of the perceived limitations.

THEORETICAL

An explanation is provided for those who like to know a little more about such things. To make the theory aspects clearer, please refer to the circuit given in Fig 1.

The 555 is used as a monostable oscillator and the test capacitor, which we will call C_x, is first charged and then discharged, and the meter indicates the average discharge current. The formula is

$$I_{ave} = \frac{V \times C_x}{(R_b + 2R_b) \times C1} \times K$$

where V is the voltage to which C_x is charged and K is a constant, depending on the charge and discharge time of the 555 circuit, including the contribution made by the IC internal structure as well as the external resistance ratios and C1.

CONSTRUCTION

Various approaches are possible but using the contents of the junk box involves little or no cost at all.

Front panel

The specification calls for a 50μA meter mounted on a plastic box. The bottom range reads 50pF full scale. A switch provides extra ranges. There are five rotary switch positions, extending the ranges by a decade factor each time.

Range extender

We now have, at position 5 of the switch, a full-scale direct reading of 0.5μF. At this point, the meter needle starts to flicker visibly at the 555 oscillator frequency. In order to extend the range further, a ×10 shunt is paralleled across the basic 50μA meter movement. This extends *all ranges* by a ×10 factor, and the maximum becomes 5μF. As the meter flicker is unacceptable, an auxiliary switch also brings in a 100μF smoothing capacitor across the meter movement.

The applied voltage is about 2V on the terminals, which is unlikely to harm even the most delicate components.

Table 1. Components list. Numbered components not listed below are for text reference only	
R1	1k PC-mounting
R2	10k PC-mounting
R3	50k PC-mounting
IC1	555 timer
S1	SPST switch
S3	2-pole 5-way rotary switch

Construction details

Short internal wiring is preferable and the IC PCB is mounted right at the test terminals. Additionally, the 555 is mounted in a socket for peace of mind, allowing easy replacement should this ever be necessary.

Calibration

Use a good 0.047μF capacitor on range 3 to set the 50kΩ calibrator to produce a reading of 47μA. Then set the ×10 range (using the appropriate switch settings) with its preset to read 5μF full scale. Next adjust the 50pF and 500pF controls. You are now able to see the difference between 2.7pF and 3.3pF accurately.

Low-cost operation

The unit has a small mains power supply and needs a secondary winding able to provide anything between 9V and 14V DC to the regulator input. If you do not like to wire mains-operated accessories, a 9V alkaline battery can be used but do not omit the voltage regulator.

CONCLUSION

There is great satisfaction to be had in building up reliable and accurate test equipment. Spend some time on the cosmetics; try to obtain matching knobs, give consideration to using brass flathead machine screws where they appear on the outside, and if silk-screening is out of the question you can use Dymotape® labels. Rubber stick-on feet ensure the case does not scratch anything and stops it sliding off the bench.

With a few known 2% calibration capacitors used from time to time as reference sources, this meter has been found to provide consistent and quite accurate measurements. The prototype has been in regular use for several months and performs consistently well.

A simple converter for the 10m satellite band

Many amateur radio satellites pass by every day. They re-transmit signals from amateur radio ground stations and each satellite generates at least one beacon signal. The signals from the amateur radio satellites in the 29.3 to 29.5MHz segment of the 10m band are usually CW or SSB. They are converted by this simple unit to the 80m band for reception with receivers such as the Alivo, which is described elsewhere in this book. The converter will work with any receiver that can tune from 3.5 to 3.7MHz and which can receive CW and SSB signals.

WHAT'S UP THERE?

Several Russian satellites make regular and frequent daily passes, some 19 to 22 passes each day in polar or low earth orbit (LEO) at altitudes around 1000 and 2000km. Two satellites may pass by at the same time! The beacon signals from these satellites are consistent and easy to find.

This converter is sensitive and a preamplifier is not necessary. Satellite beacon signals can be heard in the noise for a short period as a satellite approaches from over the horizon at the predicted time. The signals are well above the noise for most of a pass, which can be from only a few minutes to 30 minutes, until the satellite disappears back into the noise and beyond the horizon, to reappear again some 100 or so minutes later.

WHAT DO THEY DO?

The satellites transmit *downlink* signals, re-transmitted from amateur stations on the ground, and operating in *Mode A* (2m uplink to the satellite, 10m downlink from it).

This simple receiving set-up is an introduction to satellite communications. It can lead to the study of satellite orbital and celestial mechanics and to the decoding of satellite telemetry signals. Information about the status of the satellite and its hardware is sent in the beacon transmission and is available to any listener. It is easily logged and decoded. The trends of several of the satellite's parameters can be followed.

There are many satellite passes, so the chance of hearing a satellite by just listening at random times is reasonably high. Computer programs for accurately predicting the times of satellite passes are freely available. More about those later.

Listening to amateur radio satellites leads to many exciting self-training activities: building and operating the receiving equipment; listening to beacon transmissions from satellites; monitoring amateur stations working through the satellite; displaying the current position of the satellite on a map on a computer screen; predicting the satellite's orbit by computer; recording

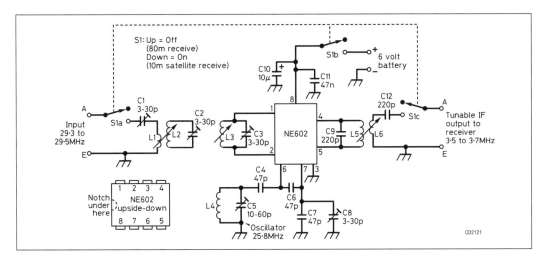

Fig 1. The circuit of the 10m converter

and decoding the telemetry information to follow changes in the spacecraft's status, and much more. All these activities develop from, and are supported by, the signals received by a simple receiver and its attached converter. These activities are the things that school science and technology projects are made of – investigatory work requiring discipline, organisation and record-keeping, essential to budding engineers and scientists.

THE CIRCUIT

This simple converter circuit, shown in Fig 1, is based around the popular NE602 integrated circuit, the same amplifier-mixer-oscillator IC as used in the Alivo receiver. A 6V battery provides the energy supply.

There is only one control – a toggle switch. It acts as a band switch to switch from the 80m band to the 10m satellite band. Three poles of a miniature four-pole changeover toggle switch are used. One pole acts as the battery on/off switch (S1b) with two poles (S1a and S1c) switching the converter in and out of circuit. Remember to switch the unit off, when not in use, to conserve the batteries!

THE OUTPUT CIRCUIT

Starting at the output, IC pins 4 and 5 feed a tuned circuit, broadly tuned to 3.5 to 3.7MHz. A link coil, L6, feeds the output signal through a short length of twin-lead or twisted pair to the receiver antenna and earth terminals. Capacitor C12 is for matching purposes; its inclusion produces a definite peak when using the peak control in the connected Alivo receiver. Coaxial cable is not necessary for the link to the receiver.

L5 is 30 turns of 26-gauge enamelled wire on a 6mm slug-tuned former. The same wire gauge is used for all coils in this converter. The link coil L6 is 6 turns wound over the bottom of L3. An untuned toroid transformer is an alternative for this output arrangement. The tuned circuit was adopted to give some added IF selectivity in the expectation of the possible appearance of out-of-band unwanted signals. This has not been a problem.

THE OSCILLATOR

IC pins 6 and 7 connect to the oscillator section. This is a free-running LC circuit. The oscillator frequency is 25.8MHz but more about this selection later. Cheap quartz crystals are not readily available for this frequency, so a Colpitts LC oscillator has been used as shown in the diagram.

The coil L4 is 10 turns on a 6mm former. The slug has been removed from the former. For stability reasons, the coil and former are enclosed in a soft plastic foam jacket held in place by a metal saddle soldered to the PC board. An earthed metal shielding can to contain the coil completely could be used. The trimmer capacitor C8 is mounted so that it can be adjusted with a trimming tool from outside the converter unit through a hole in the box, should a slight frequency change be required.

The oscillator is surprisingly stable. During construction, soldering and mechanical work on the oscillator components puts stress on those components. After several days in a closed box, the oscillator frequency settles down and stabilises. It wanders very slowly in the range about 5kHz above and below 25.8MHz during each day, irrespective of whether it is left switched on, or only switched on when it is required. The unit should not be left in direct sunlight. No special selection of components has been made to improve the frequency stability.

The oscillator frequency varies by less than 100Hz during a long satellite pass. This has been found to be quite acceptable. The Doppler frequency shift observed on the incoming signals is very much more than this so any steady long-term oscillator drift in such a simple unit is tolerable. The receiver tuning control is swung across the satellite downlink band during a pass looking for signals and continuous tuning adjustments are made to correct for Doppler shift, so the absolute frequency at any time is not of major importance. The initial reception of the satellite beacon signal gives an indication of any dial error for that particular pass.

THE INPUT CIRCUIT

The twin inductors L2 and L3 with their associated capacitors form the input tuned circuit, tuned to the 29.3 to 29.5MHz satellite band. This band-pass filter configuration was chosen to minimise strong local 80m signals. The circuit shown has made the occurrence of breakthrough rare.

There can still be a problem at times with exceptionally strong local 80m stations. Traps for these signals could be added to the converter but the added complexity is not warranted. Upper sideband is used on the satellite downlinks. This is different from the 80m band, where lower sideband is in general use, so it is easy to identify any LSB signal which may be heard as IF breakthrough. Switching to 80m reception will confirm if the signal is on the 80m band. The problem is insignificant.

The coils L2 and L3 are identical – each is 12 turns wound on two separate 6mm formers. An input link L1 of three turns is wound over the bottom of L2. The coil formers for L2 and L3 are mounted side-by-side, in parallel, about 20mm apart, centre-to-centre. There is sufficient coupling between them and no gimmick capacitor or other coupling supplementation is required.

CONSTRUCTION

The unit is built in a plastic box about 150mm wide by 55mm high and 90mm deep, available from many sources. A smaller box could be used. A completely sealed box is needed to keep air movements away from the oscillator components and to increase thermal delay in the interests of oscillator stability.

The IC is glued upside down to a piece of PC board and wiring is by the 'dead-bug' method. Tag strips support other components and the coil formers. The PC board is bolted to the lid of the box, which is then positioned upside down, so the lid becomes the base.

The two critical things about the wiring are to keep the signal input and output leads well separated (to minimise break-through of 80m signals), and to make sure that the leads to the oscillator components are rigid (to IC pins 6 and 7). Use the *outside* sections of the four-pole toggle switch for S1a and S1c to improve isolation of the input and output signal leads.

As with all battery-powered devices, be certain that the batteries are connected with the correct polarity!

THE FREQUENCIES INVOLVED

With the 25.8MHz oscillator in the converter and the receiver tuned to 3.5MHz, reception will be at 25.8 + 3.5 = 29.3MHz. Tuning 200kHz higher to 3.7MHz, it will receive at 29.5MHz. These two frequencies can be useful in calibrating the tuning scale.

Other oscillator frequencies could be chosen to give other tuning ranges on the 80m band and would be just as effective. If loud local 80m break-through is a persistent problem at the satellite frequencies of interest to you, the oscillator could be shifted 100kHz lower to 25.7MHz to make the satellite band read from 3.6 to 3.8MHz on the receiver dial.

SETTING-UP

Check the circuit and your wiring. Connect the converter to the receiver and switch on. Coil L5 is peaked for maximum noise at 3.6MHz – the middle of the reception range.

The oscillator trimmer capacitor C8 is set to a half-way position. The oscillator tuning capacitor C5 is then adjusted to set the oscillator to 25.8MHz. Listen at that frequency on a communications receiver with a digital readout and adjust C5 until you can hear the oscillator carrier in the centre of the pass-band. The trimmer capacitor C8 is used for fine adjustment through a hole in the box. Put a piece of tape over the hole to keep the box sealed.

If you don't have access to a communications receiver, try a push-button FM broadcast receiver; even a push-button FM car radio is suitable. Hold the converter near the FM receiver antenna. Listen at 103.2 MHz. This is the 4th harmonic of 25.8MHz. You should hear the FM noise decrease with a swish as the converter oscillator is adjusted. Set the converter oscillator to full quieting on the FM receiver; the correct position is easy to identify.

Due to slight variations in the values of the fixed capacitors used, it may be necessary to add or remove a turn to your coil to get the oscillator to the correct frequency.

Attach the antenna lead! The three input tuned circuit trimmer capacitors C1, C2 and C3 are all set to a halfway position. The receiver is set to 3.6MHz for this adjustment, with the incoming 'noise' peaked at 29.4MHz – the centre of the satellite band. The slugs in the formers L2 and L3 are adjusted for maximum received noise. The trimmers C2 and C3 are for fine adjustment – the noise peaks are sharp.

Be aware that if there are more turns on the coils L2 and L3 than necessary; it may be possible to set the slugs too far into those coils and obtain a noise peak at 25.8 – 3.6 = 22.2MHz.

The input capacitor C1 is then slowly moved to its minimum capacitance position. The noise heard should decrease. Advancing the trimmer back towards maximum capacitance will increase the noise level. The noise will be found to increase and level off. Set the trimmer to the point where the noise appears to first reach this plateau. Too much capacitance will be an invitation for 80m breakthrough to be a nuisance. This adjustment will depend upon the characteristics of the connected antenna. The setting is not critical.

Now listen for a satellite beacon signal at the frequencies given below. When a satellite beacon signal has been identified, it can be used to carefully adjust the input trimmers C2 and C3 even more accurately.

Once correctly set up, there is little need for any further changes to all these settings.

SATELLITE RECEPTION

The International Radio Regulations of the International Telecommunication Union show that the whole 10m band (28 to 29.7MHz) is available to the Amateur Satellite Service. Amateur radio band planning has assigned the band segment 29.3 to 29.5MHz to exclusive amateur satellite use.

The frequencies for each of the RS series of satellites are given in the amateur radio literature. The details you need to get started with each satellite are:

RS-10: Beacon 29.357 or 29.403MHz; downlink 29.360 to 29.400MHz.
RS-12: Beacon 29.408 or 29.454MHz; downlink 29.410 to 29.450MHz.
RS-15: Beacon 29.3525 or 29.3987MHz; downlink 29.354 to 29.394MHz.

The first beacon frequency listed for each satellite seems to be the one in general use.

These frequencies show our band of interest to be 100kHz wide and centred on 29.4MHz. So it is appropriate to use the RS-12 beacon at 29.408MHz to peak up the converter's input tuned circuits C2 and C3 finally for optimum performance.

The RS-16 satellite is now in orbit and adds six or seven more satellite passes each day for the keen listeners to the 10m satellite band. Its beacon and transponder output frequencies are given as being similar to RS-12.

UNDERSTANDING THE TELEMETRY

The beacon gives repeats of "CQ de..." and its RS number, followed by text and telemetry information, in Morse code. For example, the telemetry from

RS-10 is a frame of 16 four-character words. By carefully noting and processing this detail, one can obtain an interesting appreciation of the goings-on inside the spacecraft. There are 16 monitoring points and 16 measurements are sent in this repeating cycle.

A book listed in the references below contains information about how to interpret the RS-10 telemetry information. RS-12 is said to be a copy of RS-10. The RS-15 telemetry information is different. It is given in a document downloadable from the AMSAT Internet website, and the details follow below.

Most satellite beacon transmissions on this band are in CW. The Morse code speed may be too fast for the novice. There is a simple way to overcome this. Use a two-speed tape-recorder. Receive a beacon transmission using a high-pitch CW tone and record it at the fast tape speed. Play the recording back later at a slower tape speed!

SUITABLE RECEIVING ANTENNAS

A simple length of wire is a basic antenna for HF reception. So the same random length of wire used for many experiments, about 20m long, strung out the window, is used for reception of these satellite signals. It is probably too long for the purpose but it works! A half-wave dipole antenna cut to length for the 10m band, positioned up clear of obstacles and centre-fed with coaxial cable, or a vertical 27MHz CB antenna, may be better receiving antennas. Experiment! There are many good antenna textbooks that can assist.

AN AUDIO FILTER

An audio filter is a useful accessory for any receiver when seeking narrow-band signals in noisy conditions. Weak signals from approaching satellites, rising up from the noise at a distance of more than 4000km, make a supplementary audio filter useful but not essential. Such a filter is described in another article in this book.

At times it is possible to hear a 10m satellite beacon signal at very long distances away, well beyond the horizon. This is due to the characteristics of HF propagation. An audio filter is useful to help to identify these signals in the noise.

WHEN TO LISTEN?

Three ways are suggested to find the times of arrival of the satellites. The first is to listen at random times over, say, a two-hour period. These satellites have an orbital time of some 105 to 128 minutes. With some 20 passes each day each of up to 30 minutes' duration, there is a chance with random listening that you will hear signals!

A second way is to ask a local amateur radio satellite enthusiast to provide you with a computer printout of predictions for the 10m band satellites for the next few days for your location. This will give you the times to listen and a lot of other details about the position where the satellite comes up over your horizon and where and when it disappears.

A third and preferred method is to set up a personal computer and a suitable computer program to produce this orbital data for yourself at the time you want it. The computer is a very useful tool to support many amateur radio activities. A computer satellite prediction program will show when a satellite of interest will be visible from a specified location. This may be by graphical means with a map, or by a table of times and directions which can be printed out. Such programs give the AOS (acquisition of signal), the time when a satellite clears the horizon to come within your line of sight, and the LOS (loss of signal), the time when the satellite passes dips below the horizon.

The time you first hear the beacon signal from an approaching satellite depends on several things, including hills and other obstacles that may obscure your view of the computed horizon. There are occasional satellite passes in which a satellite, at a range of more than 4000km, rises only a degree or so above the computed horizon and for only a few minutes. These distant passes may be obscured by high hills on the visual horizon in that direction. Reception of the satellite signals with any receiving equipment in those circumstances, if it is possible, will probably be weak and noisy.

There are several good computer programs for satellite prediction available by download from the Internet. Some are free; others require a small donation. Start at the AMSAT website: www.amsat.org. AMSAT is a non-profit-making organisation of volunteers promoting and supporting amateur radio satellite activity.

Some programs have excellent graphics but may require as a minimum a 386 or better computer. A good explanatory document accompanies some programs.

To remain current, all satellite computer programs need regular updating of the orbit information by entering new *Keplerian elements*. These are explained in simple terms in documents available for download from the AMSAT website.

Regular Keplerian element update information is available from amateur radio packet bulletin boards and from various WWW sources. If you have electronic mail facilities, you can put your details forward at the AMSAT website to get on a free Internet mailing list to bring Kepler updates direct to your computer by e-mail message each week. This is an AMSAT service to promote interest in satellite activity. Most satellite programs have procedures to accept Kepler updates automatically from messages in a two-line format, so tedious typing and error-checking are not necessary.

AMSAT needs our financial support for the hardware in the sky and for the resources of technical information so readily provided. Details about the various AMSAT groups are on the website and it has many downloadable documents for the budding satellite enthusiast. One gives a very full description of the RS series of satellites.

With a computer program correctly installed for your location, and with the appropriate current Kepler elements, the position and ground track of several satellites, such as the Russian RS series, the Space Shuttle, the Hubble Space Telescope and the Gamma Ray Observatory, can be obtained. Many of these satellites can be observed visually with the naked eye in the dark of the evening as they pass by, weather and other conditions permitting. With an amateur radio licence and a VHF transceiver, you can talk to radio amateurs on the International Space Station!

MODIFICATIONS?

An Alivo receiver (described elsewhere in this book) could be modified and built as a dedicated receiver for the 10m satellite band. This idea was considered and rejected in favour of the converter approach. The converter can be made fixed-tuned for the relatively narrow satellite segment and the receiver is still available for 80m reception. Building this converter is easier than building an Alivo receiver.

IN CONCLUSION

Satellites are a very interesting aspect of amateur radio and it is very easy to get hooked. Just build a simple 80m receiver with a satellite converter. It is a great thrill to receive satellite signals using equipment that you yourself have built. This can lead to so many other things to investigate. Satellite pass times soon dictate your daily routine. A challenge indeed! Give it a go, and have fun – there is much experimenting to do!

REFERENCES

[1] 'The Alivo receiver', Fred Johnson, ZL2AMJ, *Break-In* September 1996, pp4–12.
[2] *The ARRL Satellite Anthology*, 3rd edn, ARRL, 1994, pp114–119 inclusive. This is the best of *QST* articles on amateur satellite operation and hardware. Telemetry decoding for RS-10 is on pp115 and 116.

Noise-reduction circuits

Many people have transceivers with digital signal processing (DSP) inside them. Here is a simpler approach, based on an analogue integrated circuit. The prototype used some surface-mount devices (SMDs) but this is not necessary to enjoy the benefits of this little circuit.

DSP

The test reports on digital signal processors promise miracles; many people feel that the noise is reduced in the noise reduction mode but so is the readability!

The Analog Devices SSM2000 IC [1], works on the HUSH principle, developed and patented by Rocktron Corporation. Although designed for the hi-fi market, it has found an application in amateur radio which is described here.

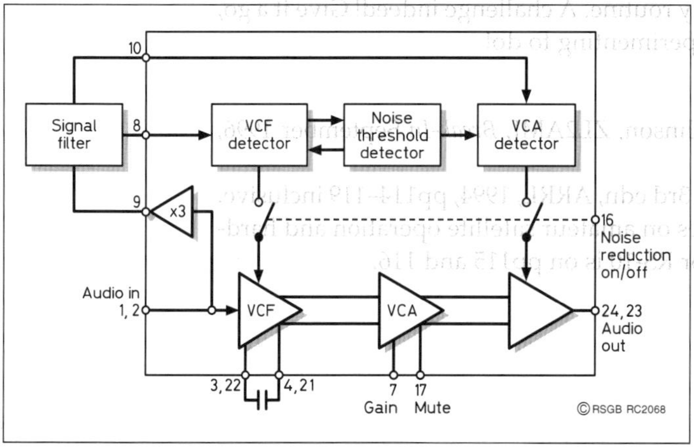

HOW IT WORKS

The HUSH system can distinguish between signal and noise because the volume and frequency spectrum of speech or music continually changes; by contrast, the noise amplitude and frequency spectra remain relatively constant.

Following the block diagram in Fig 1, the audio signals

Fig 1. Block diagram of one channel of the Analog Devices SSM2000 IC

(only one channel of the two required for stereo sound applications is described) are processed to extract information concerning the frequency distribution and amplitude of both signal and noise, passing through low-pass voltage-controlled filters (VCFs) and then through voltage-controlled amplifiers (VCAs). In contrast to digital processes, VCFs and VCAs have low distortion and add negligible noise of their own. The cut-off of the VCF and the gain of the VCA can be set as required by the application. With control signals derived by a proprietary algorithm and applied to both VCF and VCA, a noise reduction of up to 25dB can be achieved.

THE APPLICATION

Looking at the diagram in Fig 2, the signal is applied to pins 1 and 2. It comes out of pin 9, amplified by a factor of three, from where it is fed unaltered to the VCA detector (pin 10) and through a three-pole high-pass filter to the VCF detector (pin 8). Fig 3 shows the two frequency responses, optimised here for hi-fi music; for narrow-band voice signals it would be appropriate to move the frequency response of the three-pole filter downwards by making the capacitors about five times bigger (22nF to 100nF, 22nF to 100nF and 2.2nF to 10nF respectively).

The time constant for the decay of the control voltage to the VCFs is set by the 1µF capacitor on pin 11. The noise threshold is 'held' on the 220nF capacitor on pin 15. The control voltage to the VCA decays according to the value of the capacitor on pin 12. The lower cut-off frequencies of

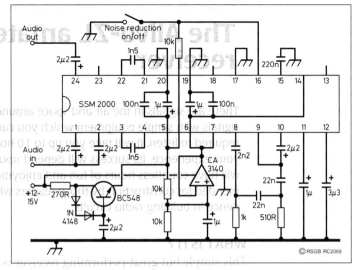

the VCFs are determined by the 1.5nF capacitors between pins 3 and 4, 21 and 22.

The supply voltage on pin 5 is decoupled by the BC548 emitter follower. The op-amp holds pins 6, 7 and 14 on half the supply voltage. The switch on pin 16 allows the noise suppression to be disabled.

Fig 2. The noise reduction circuit, for use in a communications receiver

RESULTS

The unit was built on a double-sided 52 by 33mm PCB. Except for the DIL ICs, SMD components were used throughout. Through-connections are by way of the terminal pins and the pins of the DIL sockets, which are soldered on both sides. The two 1.5nF capacitors were soldered directly to the IC.

There are no adjustments. As an audio input of 0.3V is recommended, the noise suppressor was installed between the product detector and the volume control of the receiver, an old Drake R4-C. The product detector output is fed to both the L and R inputs of the SSM2000. The sum of these inputs may be applied to pin 9 of the VCA and VCF controllers. The output to the volume control is taken from one channel only.

The manufacturer claims a noise reduction of up to 25dB in hi-fi systems. This cannot be obtained with speech, but the performance is very pleasing as it is achieved without audible distortion of the wanted signal.

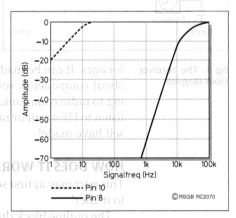

Fig 3. Low-frequency roll-off to pin 8 through the 3-pole filter in Fig 3. Also shown is the signal on pin 10, which does not go through the filter

REFERENCES

[1] A 16-page data sheet for the SSM2000 can be downloaded from http://www.analog.com.

The Alivo-ZL amateur radio receiver

There are signals in the air and space around you. You can listen to these signals with simple equipment which you can make yourself. No licence is required to listen. It will take you up to 10 hours to make, depending upon your experience. Its success will depend upon the care you take, and you will have countless hours of fun and enjoyment from listening. It is recommended for construction by those readers who have a fair degree of experience in building radio circuits

WHAT IS IT?

This simple but great-performing receiver is safe and easy to build by constructors of any age – a 'hands-on' introduction to amateur radio.

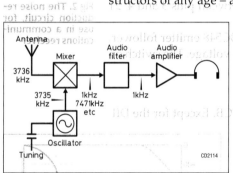

Fig 1. The receiver block diagram

Battery-powered, it covers the popular amateur radio 80m band of 3.5 to 3.8MHz. It brings the excitement of listening to amateur radio signals, live, as they happen. Phone signals, both single-sideband and amplitude-modulated, and Morse code signals (continuous wave – CW) can be received.

Its features are great sensitivity, simplicity, low cost, presentable in appearance and comfortable to use. It requires light-weight stereo headphones and a length of wire as an antenna. The batteries last for ages. It can be made at home using common hand tools, while learning about components, soldering, wiring, tuned circuits, antennas and listening to radio communication on-air. A great way to learn about radio and to listen to HF amateur radio communications – with a receiver that you yourself have made!

HOW DOES IT WORK?

This part may at first seem complicated – skip it if you wish and come back to it later!

The outline block diagram is in Fig 1, and is of a *direct-conversion receiver*. A steady carrier signal at a frequency of (say) 3.736MHz (or 3736kHz) is fed to a mixer. A local oscillator at 3735kHz will beat with this signal and produce various output frequencies, such as the difference between the two signals, 1kHz, and the sum, 7471kHz.

The output from the mixer is passed to an audio filter. The 1kHz signal passes through and is amplified by the audio amplifier. The 1kHz sound is heard in the headphones. Changing the frequency of the oscillator with the tuning control – as the receiver is tuned across the band – will change the

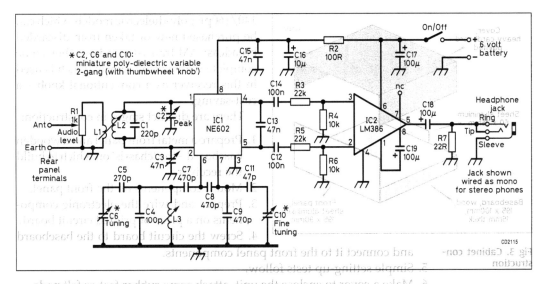

Fig 2. The circuit diagram of the complete receiver

pitch of the audio signal you hear. SSB (single-sideband (speech)) and CW (continuous wave (Morse code)) signals can be tuned in by careful adjustment of the tuning control.

THE CIRCUIT DIAGRAM

The circuit is shown in Fig 2. Integrated circuit IC1 acts first as a radio-frequency amplifier and then as a mixer, with L3 and various capacitors forming the oscillator. It is followed by IC2, an audio amplifier.

The potentiometer R1 in the receiver input circuit is an attenuator. At one end of its travel, it earths the antenna. At the other end it feeds the incoming signal from the antenna direct to the primary coil L1, coupling from there into the tuned circuit L2 and C1 (with C2) and to pins 1 and 2 of IC1. So R1 controls the level of signal through the receiver and the loudness of the audio heard at the headphones. C2 is peaked for maximum signal.

The audio filter is between pins 4 and 5 of IC1 and pins 3 and 2 of IC2, with the audio output lead from pin 5 of IC2 to the headphone jack socket. A single capacitor C19 between pins 1 and 8 of IC2 sets the audio gain. The positive lead of the 6V battery connects through the on/off switch to pin 6 of IC2 and to pin 8 of IC1. Pin 3 of IC1 goes to earth and pin 4 of IC2 is also earthed, completing the DC path through each IC. The battery negative lead is also connected to earth. 'Earth' and 'chassis' mean the same thing here – the common metal front panel and the copper of the circuit board.

Pin 7 of IC2 is not connected (NC) to anything.

CONSTRUCTION

All components for this receiver can be purchased new from electronics parts suppliers. Some may be found in 'junk boxes' but be sure to use only good-quality items. Many other parts can be substituted.

Three variable capacitors are required, for the 'tuning' (C6) for the 'fine tuning' (C10) and for the 'peak' control (C2). These are miniature two-section

127

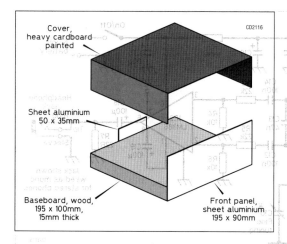

Cover,
heavy cardboard
painted

CD2116

Sheet aluminium
50 x 35mm

Baseboard, wood,
195 x 100mm,
15mm thick

Front panel,
sheet aluminium
195 x 90mm

Fig 3. Cabinet construction

140/160pF poly-dielectric models which can be purchased new or taken from discarded broadcast AM transistor radios. They come complete with a thumbwheel which is used in this receiver as a conventional knob – a cost-saving!

There are distinct stages in construction:

1. Prepare a metal front panel and a wooden baseboard – the 'chassis' on which to build the receiver.
2. Mount components on the front panel.
3. Prepare and wire the electronic components on a piece of printed circuit board.
4. Screw the circuit board to the baseboard and connect it to the front panel components.
5. Simple setting-up tests follow.
6. Make a cover to enclose the unit, attach some rubber feet or felt pads.

The receiver will then be complete and ready for service.

The box

The design shown in Fig 3 is cheap to duplicate. It is easy for anyone to build with a minimum of tools and there is still access to every component for experimenting.

The finished receiver must look the part, so take time preparing the front panel. The panel is a piece of flat sheet aluminium, 195mm × 90mm. The thickness can be about 1.2mm (18 gauge).

The front panel design of Fig 4 can be photocopied and enlarged. Stick one copy to the front panel (use a few dabs of Pritt Stik® or similar) and use it as a drilling template. For a professional appearance, a second copy can be laminated in clear plastic and glued on the finished front panel.

Carefully check the physical position of each component mounted on the front panel. Check the size of hole required for each item by measuring from the component itself – before you drill! Do all the drilling and filing and finish

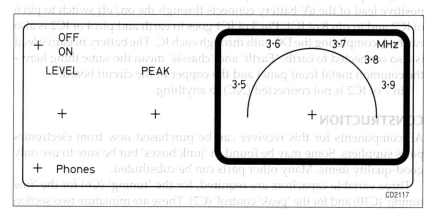

OFF
ON

LEVEL   PEAK

+   +

+ Phones

3·6   3·7   MHz
3·8
3·5
3·9

+

CD2117

Fig 4. Front panel design – actual size is 195mm × 90mm

the front panel metalwork before any components are mounted. Take care and do a neat job.

Mount the three variable capacitors with care. You may need to space them back from the panel with a spacer. Failure to do this could result in the capacitor mounting bolts projecting too far into the interior of the capacitor and damaging it. Shape a paperclip into a 'U' as shown in Fig 5. Place it between the panel and the capacitor.

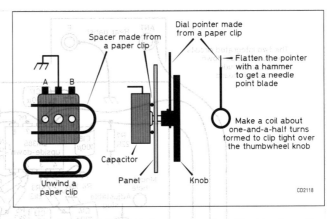

Fig 5. Capacitor mounting and fabrication of dial pointer

Check the length of your mounting bolts and the thickness of your panel. You may need a spacer that is thicker or you may not need one at all.

The two built-in trimmer capacitors inside each variable capacitor should be set to mid-value. On the back face of each capacitor are two screws. Rotate each so that the trimmer capacitor (seen through the plastic) is half-meshed (at about mid-capacitance). This is important to get the correct tuning range for the receiver during the setting-up.

The baseboard is a piece of wood 195mm × 100mm and about 15mm thick. Select a piece that is clean with neat edges and cut it to size. Chipboard or MDF is ideal. Paint it with several coats. Screw the front panel to the baseboard with self-tapping screws. The antenna and earth terminals are mounted on an aluminium panel, 50mm × 35mm, screwed on the back edge of the baseboard.

The tuning dial

The main dial operates the oscillator tuning capacitor, C6. The pointer is made by unwinding a wire 'glide' type paperclip, and making a needle as shown in Fig 5. One end is formed into a spring to grip the hub of the thumbwheel knob (which comes with the tuning capacitor). This makes the needle adjustment easy. The other end is flattened between a hammer and a hard surface to make a precision blade pointer!

Set the needle to point to zero when C6 is fully anti-clockwise.

Preparing the circuit board

Start with a piece of blank printed circuit board stock, 75mm × 50mm, as shown in Fig 6. This is insulating material with a sheet of copper bonded to one side.

File all the edges and corners to make them clean and smooth. Drill the mounting holes as shown. Polish the copper with a scrap of steel wool to make it clean and bright. Hold the board by its edges to keep it free from finger-marks. Spray it with clear polyurethane varnish from a spray can, a very thin covering – a quick swiping flick of paint spray. Leave it for 24 hours to dry hard. You can solder through this paint layer and it will keep the board looking bright.

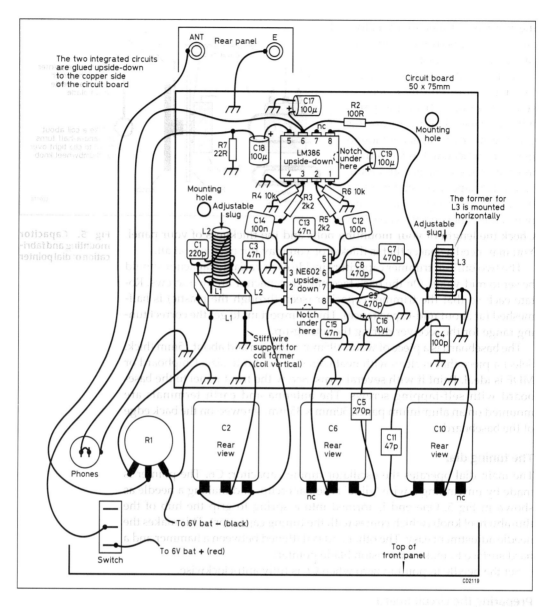

The two integrated circuits are glued upside-down to the copper side of the circuit board

Fig 6. Component connection diagram for the complete receiver

Mounting the integrated circuits

The ICs lie on their backs on the board in 'dead-bug' fashion. They must be spaced well apart so that other components can be positioned between them. The positions of many components on the board are dictated by the pin numbering of the integrated circuits. Remember to keep leads short and to place bypass capacitors close to the pins. You have very little choice in component placement and will have to place them as generally shown in Fig 6.

Scrape the paint off the circuit board at the places where the ICs are to be positioned, exposing the copper. Put a blob of glue at each of these places on the bright copper. Then position each IC in turn, legs in the air, on each blob of glue. Check the type number of each IC before placing it (you cannot read the label after it is positioned!). Use a magnifying glass. Be very sure that you position each IC so that the location 'notch' identifying the start of the pin numbering is correctly aligned.

Put the board aside until the glue has set. It is then ready for wiring.

Wiring the board

Cut the component leads to length and shape them, fixing each component into position by soldering, one at a time. Use a fine-tipped soldering iron. Keep leads short. Be neat. You can solder through the paint on the board to get the various 'earths' (to the copper) for the components.

Soldering for the first time? Try some other scrap components for practice first. Read about soldering techniques in amateur radio reference books.

Take time and be patient when wiring the board. Be careful when using the soldering iron and avoid overheating components. Heat the joint, not the solder. Apply the solder to the joint when the joint is at the right heat. Be sparing with the solder. Clean the tip of the iron frequently, grab it with a clean rag folded several times to avoid burning yourself.

Each IC has one pin which is earthed. Use scrap wire discarded from a component lead to connect the earth pin of each IC to the copper board.

Do not twist or wrap the component leads around the IC pins. A

Table 1. Components list

Value	Quantity	Notes
Resistors		
22R	1	
100R	1	
2k2	2	
10k	2	
1k	1	panel-mount potentiometer, R1
All resistors ¼W or smaller		
Capacitors		
47p	1	
100p	3	
220p	1	
270p	1	
470p	6	
10µ	1	electrolytic, 10V or more
100µ	3	electrolytic, 10V or more
60–160p	3	polyvaricon variable capacitors, C2, C6 and C10
All capacitors (ceramic OK) low voltage		
Integrated circuits		
NE602	1	
LM386	1	
Additional items		
Coil formers, 7mm dia with 9mm square base (or very similar): 2 needed		
Slugs to fit above formers: 2 needed		
Battery-holder for 4 AA cells		
Clip & lead for battery-holder		
Phone jack, 3.5mm panel mount, stereo		
Knob, small, for potentiometer		
Terminal red (for antenna)		
Terminal black (for earth)		
Switch, miniature SPST		
Circuit board, blank, 75mm × 50mm		
Baseboard, wood, 195mm × 100mm		
Panel, 1.2mm aluminium, 195mm × 90mm		
Panel, 1.2mm aluminium, 50mm × 35mm		
Bolts to mount tuning capacitors, M2.5 × 6mm: 6 needed		
Screws, self-tapping, 4g × 10mm: 12 needed		
Paperclips (glides): 4 needed		
Wire for coils, 0.4mm (26SWG enamelled): about 2m needed		
Sundries		
Cored solder		
Small pin nails & rubber band (if required)		
Hook-up wire, tinned-copper wire etc, scraps, as required		

component lead has to only touch the pin and the solder will do the rest. Bend the component lead so that a 1mm (say) length of the lead is making contact with the pin, then solder it.

Resistors can be soldered in place either way around. The same applies to capacitors – except for electrolytic capacitors, which must be correctly polarised. The diagrams show the positive lead for each electrolytic – there are four of them: C16, C17, C18 and C19. The component itself usually has the negative lead marked, and it is usually slightly shorter (on new components).

Note that there are three 'mid-air' joints between components – the junctions between C14 and R3, C12 and R5, C18 and R7. Ensure that these joints are clear of the board surface. You may find it more convenient to make these joints first, then shape and attach the assembled components to the ICs. Add R2 last, wired so that it takes the short path across the top of the two ICs. Strip two pieces of insulation off some hook-up wire and use them as insulated sleeving on each end of R2. Push them on the resistor fly-leads as insulation.

Winding the coils

There are three coils to wind on two coil formers. There are many different types and sizes of coil-former and slug, so it is impossible to provide winding details for more than one type here. The following detail applies to coil formers 7mm in diameter with a 6mm diameter slug, and 16mm long. These formers should be obtained to make the construction and setting-up easy.

Start with L2 and L3 which are each made by winding 30 turns of 0.4mm (26 or 30SWG) enamelled copper wire on two separate slug-tuned formers. Other wire of a similar thickness can be used. Scrape the enamel from about 5mm at the end of the wire to expose the bare copper. Twist it around a wire tie-point on the coil former. Solder the joint. Wind the coil for the full 30 turns. Bring the wire down the outside of the coil and repeat the clean and soldering task to another tie-point, then cut off the surplus.

L1 is 6 turns of the same wire scramble-wound over the 'cold' (ie earthy) end of L2 (the end that goes to C3). The coil end nearest to the tie-points should be adopted as the 'earth' or 'cold' end of the coil. The coils L1 and L2 must be isolated – there must not be a direct short-circuit between them. The coil former of L1 and L2 is mounted vertically by soldering the L1 earth coil-former pin to a scrap piece of heavy-gauge stiff copper wire. That wire is then soldered to the board as shown in Fig 6. Keep the two leads from L2 to IC1 short.

Coil L3 is mounted in a similar way – using stiff wire as stand-offs to hold the coil horizontal. It must be mounted so that it is firm and rigid, without any loose movement. Brace it with a stay, too. The frequency stability of your receiver depends upon this. Run some glue along L3 to keep the turns in place. The adjustable ferrite cores ('slugs') inside the coil formers of L1/L2 and L3 will be positioned at various places through their tuning lengths during setting-up. For this process, you will need a plastic trimming tool. If you haven't got one, make one up from an old plastic knitting needle. File one end to shape so that it fits neatly into the hole or slot in the slug. Be careful with these adjustments and do not use any force. Slugs are brittle! Later, when you have the slugs in the coil formers L1/L2 and L3 in their final positions, you can fix the them in place in the coil. One method is to remove the slug, drop a strand of sewing cotton down the hole, and wind the slug back into its final position. If the correct size of cotton is used, the slug will still be adjustable but firmly wedged. Take care! The slugs must not be loose inside the coils.

Check the board!

Is each component correctly placed? Are the electrolytic capacitors correctly polarised (negative and positive leads to the correct places)? Check and re-check the IC pins – are the correct ones earthed? Does each pin have the correct components attached? With a magnifying glass, examine each soldered joint. Is it a solid connection? Are there any 'dry joints'? Are there any solder bridges shorting any pins? Take time, you have to get it right!

Board mounting and panel wiring

The battery-holder goes at the left end of the baseboard (looking from the front). The battery holder is held on the baseboard with two screws, or alternatively, four small nails, two each side, bent over to form hooks, with a heavy rubber band clipped over the top.

The circuit board must be positioned so that the leads to the front panel tuning controls C2, C6 and C10 are kept relatively short. Fig 6 shows the connections.

Use stiff tinned-copper wire (0.7mm) for the front panel wiring where possible. Floppy wires could cause undesirable frequency shifts. Be sure to keep the leads to the variable capacitors short and rigid.

Lightweight stereo headphones work well with this receiver. Choose them with care – some are more sensitive than others. The phone jack automatically connects stereo phones as mono – the tip and the ring contacts are joined.

Wiring the variable capacitors

These capacitors have two sections of unequal capacitance connected to three connecting lugs. Looking at the end of the shaft as shown in Fig 5, the centre lug is the common (earth) connection with the larger capacitance section, B, being the lug clockwise from the earth lug. Different connections are used on each capacitor.

The main dial of C6 gives coarse frequency control. The presence of a signal can be identified but the actual 'tuning in' is done by the fine tuning control C10. Stick a piece of paper (one from a sticky label is suitable) to the C10 thumbwheel knob as a pointer. Set it to point horizontally to the left when the knob is fully anticlockwise.

ANTENNAS AND EARTHS

Some 20m of wire, suitably suspended, is an adequate antenna and will work well. Almost any type and length of wire can be used. Run it out of the window to a tree or building. A length of rope or cord at the distant end will act as an insulator and halyard. Suspend it high so it is not a hazard to passers-by. Keep your wire well away from power lines. Take care!

An earth connection to this receiver may improve reception (by noise reduction). Try it when you have the receiver operating. If you decide that an earth improves reception, drive a scrap length of metal water pipe into the ground as a separate earth spike for your own use. Use an electrician's 'earth-clamp' to connect the earth wire to the pipe.

GETTING IT GOING

Check all the wiring – again! Make sure that the battery switch is 'off'. Be certain that the wiring polarity for the battery connector is correct. Connect the batteries.

The setting-up adjustments

Using your adjusting tool, set the slug in L1/L2 to about one-third of its length projecting out of the bottom. Set the slug in L3 to be away from the front panel with about one-third of its length inside the winding L3. Connect the antenna. Plug in the headphones.

The time of the great event has arrived. Switch on! Some checks have to be done before you can listen to amateur radio stations. Remember to switch off when you make changes to the wiring.

Setting up the oscillator

The aim is to get the tuning capacitor C6 to cover the frequency range 3.5 to 3.9MHz and with the correct dial readings. Access to a calibrated communications receiver or transceiver for the setting-up tests can make the calibration job easy – but there are other ways too! Each receiver requires individual attention. It is not difficult – just a little patience is required. You can listen to stations to judge 'where you are' with the calibration process. Set the fine tuning capacitor C10 pointer to vertical and leave it fixed at that position until the setting-up process is complete. Find a station that can be received both on the communications receiver and on your Alivo.

Run the slug L3 through its length. You should identify an AM signal with speech (or sometimes music). You will hear howls as you tune across the signals. If you can tune it in and identify it, your task is almost done. Patience is needed but is rewarded. At night, the presence of other signals may make the task confusing but correct adjustment is still possible. The aim is to get the pointer on the C6 capacitor indicating the same frequency as the communications receiver.

The L3 slug position is sharp and critical. The slug must be finally fixed in position so that it does not move – try the cotton method described earlier. You can make another check. Use an ordinary, cheap, transistor radio that tunes across the AM broadcast band. Connect a lead from your receiver antenna terminal to the external antenna connection of the AM radio. With the latter set to about 856kHz on its dial, its oscillator will be at 1311kHz (being 856kHz plus 455kHz, its IF). The third harmonic of 1311 is 3933kHz and you should hear it whistle on your newly constructed receiver. You can use this method to check your C6 dial at other points too (3500kHz is 711kHz on the AM dial).

The objective is to get your oscillator to cover the 3.5 to 3.9MHz band with the dial setting correct. Careful adjustment of the slug in L3 should make this possible. If the frequency calibration settings are not near enough, changing C5 will alter the 'spread' and adjusting the appropriate trimmer (on the back of C6) can help too.

Setting up the input circuit

The antenna input circuit is easy to set up. With an antenna connected, noise (or an incoming signal) can be used for this setting-up test. But first, check that the 'level' control R1 is set to its minimum attenuation position – ie fully clockwise, maximum signal, the antenna going direct to L1.

The 'peak' in the noise level (or signal level) heard as the 'peak' control C3 is swung through its range will be distinct. Adjust the slug in L2 to be sure that with the 'tuning' set (on the dial) at the low end of the frequency range (below 3.5MHz), the 'peak' control shows a definite peak in noise, and will peak again with the 'tuning' set at the high end of the frequency tuning range (about 3.95MHz). Fix this slug in position too.

With an antenna connected, you should now be receiving signals – depending on the time of day!

THE COVER

A discarded cardboard carton with thick solid sides can be used to make a cover. Measure up and cut it out with shears or a sharp knife and a straight-edge, fold it, and give it a couple of coats of paint. You can choose any colour you like. Attach it to each end of the baseboard with screws. Add some self-adhesive rubber feet or felt pads to your receiver and prepare to enjoy listening!

OPERATING YOUR RECEIVER
CW (Morse code) reception

In Fig 1 we saw how signals are processed by direct-conversion from the original radio frequencies down to audio frequencies. Consider a Morse signal (CW – continuous wave) being sent on 3736kHz. This will be heard as a 1kHz note in the phones (the difference between the incoming signal and the receiver oscillator – which is at 3735kHz). As the transmitting Morse key is operated to form the elements of the Morse code characters, the 1kHz note heard by you will respond accordingly.

If your receiver oscillator is now tuned to 1kHz lower in frequency – to 3734kHz – the 1kHz note you hear will increase to 2kHz (being the difference between 3734 and 3736kHz). So as you tune your receiver across an incoming carrier signal, you will first hear a signal as a high pitch note. This note will decrease to zero and will then increase again to a high pitch as you tune across the signal. When the oscillator is at the same frequency as the incoming signal, you will hear nothing – this position is called *zero beat*.

At each side of zero beat, you can hear the signal as a 1kHz note, in this example when the oscillator is at either 3735kHz or at 3737kHz. The receiving characteristic of this receiver can be described as a *double-sideband effect*. With the oscillator set at 3735kHz, signals at 3736 and at 3734kHz will each produce a 1kHz tone in the phones (this is 3735kHz, the oscillator frequency, plus and minus 1kHz, to give you these incoming signal frequencies). This characteristic can be likened to a superheterodyne receiver with its BFO

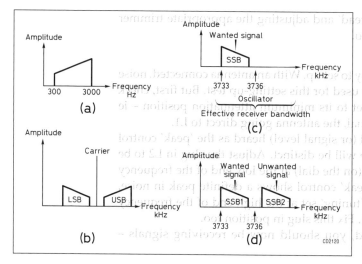

(a)

(b)

(c)

(d)

Fig 7. Spectrum diagrams to illustrate (a) the audio spectrum (idealised); (b) amplitude modulation, showing the two sidebands and the carrier; (c) single (lower) sideband, suppressed carrier; (d) receiving SSB in the presence of an unwanted signal

(beat frequency oscillator) set in the middle of its pass-band.

So the effective 'radio frequency pass-band' of the receiver is twice that of the pass-band of the audio filter shown following the mixer in Fig 1. In practice, the real limit may be twice your personal hearing 'limit of audibility' – twice the frequency of the highest frequency note that you can hear! This characteristic can be used to advantage. It gives you a choice of two 'spots' that you can choose for the reception of a CW signal – you can tune your oscillator to above or below the incoming signal and still get a suitable tone for reception of the wanted signal. One spot may have an interference advantage over the other by changing the pitch of an adjacent and unwanted signal whilst keeping the pitch of the wanted signal the same. Try it!

SSB (single-sideband) reception

To explain SSB reception, we first consider the audio (speech) input to a typical SSB transmitter to be from 300Hz to 3000Hz as shown in Fig 7(a), the diagrammatic 'wedge' symbol. The 'low end' of the wedge represents the 300Hz end, and the 'high end' of the wedge represents the top end of the speech range, 300 to 3000Hz. This diagram is for purposes of explanation only, and should not be taken to represent the voice energy distribution in actual practice. It is usually the other way around – the low notes contain most of the voice energy whilst the high-pitch notes are weak but are very necessary for speech recognition purposes.

An 'amplitude modulated' (AM) transmission can be shown as Fig 7(b) using this diagrammatic 'wedge' symbol. Here is a carrier signal with two adjacent *sidebands* – the wedges shown above and below the carrier – each indicated here as USB (upper sideband) and LSB (lower sideband). The energy of the AM signal is contained in these three components – the lower sideband, the carrier, and the upper sideband. Remember: with no input speech to the AM transmitter there are no output sidebands but the carrier continues on unchanged.

In a SSB transmission, only one sideband is radiated – both the carrier and the other sideband (of AM) are suppressed. Remember: with SSB, with no input speech, no signal at all is transmitted.

On the 80m band, it is customary for amateur stations to use the lower sideband for an SSB transmission. Amateurs use the (suppressed) carrier frequency when referring to the frequency of an SSB signal, so Fig 7(c) applies. The position for the 're-inserted carrier' (your receiver oscillator), which is needed as the reference to restore the signal during demodulation in your receiver, is shown in this diagram as 3736kHz.

Note that the LSB signal appears 'inverted'. The 300Hz component of the speech is now the higher frequency component in the transmitted signal at 3735.7kHz (3736kHz minus 300Hz). The 3000Hz component is at 3733kHz (3736kHz minus 3000Hz).

You can resolve this SSB signal by carefully setting your receiver oscillator to 3736kHz. You will hear the 300 to 3000Hz range of the transmitted audio in the phones.

However, at times you may also hear an interfering signal. If there are two quite separate but adjacent SSB signals – shown as SSB1 and SSB2 in Fig 7(d) – and you are listening to the lower frequency one, you will hear the higher frequency one as 'inverted speech'. The 3000Hz component of that higher-frequency SSB signal will be heard by you as a low-pitch audio signal and its 300Hz component as a high pitch! Fortunately this interference is almost indecipherable by the human ear. Your ear will tend to discard it as noise and will receive and listen to the 'natural-sounding' wanted signal. Of course this ear-discrimination characteristic also depends upon the relative levels of the two signals.

AM reception

Listening to an amplitude modulated (AM) signal – Fig 7(b) – on this type of receiver, requires the oscillator to be tuned very carefully to carrier 'zero beat'. A slight shift from true zero beat will show a low-speed warble effect on the audio signal that you hear. Two-thirds or more of the radiated energy of an AM transmission is in the carrier (which carries no intelligence) and the rest of the energy is divided between the two sidebands. This shows up on reception during tuning, with a very loud squeal from the beat with the carrier until zero beat is reached and with much weaker audio from the sidebands. You will soon learn how to 'tune in' SSB and other signals. This receiver is very good for demodulating the CW and SSB signals which predominate on the 80m band. AM is not frequently used today but can be received.

THE 80m BAND

Band plans for all amateur radio bands and the operating privileges for each licence grade are given in the *RSGB Yearbook*. Amateurs share these frequencies with other stations, so expect to hear signals from stations of other services. Amateur CW signals can be heard anywhere in the band, whereas amateur voice (SSB) signals are usually found within the top 200kHz of the band. Below 3.5MHz there are stations on USB with aeronautical mobile functions but their transmissions are brief and infrequent.

PERFORMANCE – AND CAN IT BE MODIFIED?

The receiver will receive all that 'communications' receivers receive. Overseas stations will be regularly copied. Many tests and experiments are possible with this receiver. Part of the fun of amateur radio is redesigning and modifying. Keen experimenters will develop improvements!

Getting started on a shoestring

Many people will tell you that amateur radio is an expensive hobby. It's true that it *can* be, but it is important to emphasise that it *need not* be too expensive to get on the air.

Many newly licensed amateurs will already have a suitable receiver, having spent an 'apprenticeship' as a listener but, once their new licence arrives, they will naturally want to transmit. The logical means might seem to be to build or buy a transmitter. However, for the beginner this is not always the easiest way to get on the air. Unless the receiver has been designed to be used with a specific, separate, transmitter, it can be quite complicated to arrange the transmitter/receiver antenna changeover switching. Another difficulty is arranging receiver muting, so that the receiver is not completely overloaded by the strength of one's own transmitted signal. At worst, damage to the receiver can occur if attention is not paid to this.

A generally more satisfactory option, especially for the beginner who is impatient to get on the air, is to buy (or build) a transceiver, rather than use 'separates' (a separate transmitter and receiver).

A *new* HF, VHF or UHF transceiver from any of the 'big three' manufacturers – Kenwood, Yaesu and Icom – will cost anything from a little under £1000 to £4000 or more, and they will give you every facility you need – and a lot you don't. A 'basic' transceiver, on the other hand, just needs to transmit and receive on one or more bands. Obviously there are several features that are essential – the AF gain (volume control) to use an extreme example – but many of the features found on modern transceivers, such as memory channels, speech processors and scanning – so-called 'bells and whistles' – while nice to have, are not absolutely necessary.

So instead of buying a new transceiver, here are some alternative suggestions of how to get on the air cheaply. But first, you must decide which band or bands you want to operate on and this will largely be dependent on your licence class. A new Class B Novice will almost certainly want a 70cm FM transceiver, but a Class A Novice – or a newly licensed Full Class A/B or A – will probably want an HF 'rig'. A new Full Class B licence holder is likely to want a 2m transceiver. For those looking for a VHF or UHF rig, the decision of whether to operate only FM or to also try SSB and/or CW also has to be taken into account.

SECONDHAND

A look through the 'Members' Ads' in the last couple of issues of *RadCom* shows that there are some real bargains out there. For an HF transceiver, how about a Yaesu FT-101ZD at £300 or an Icom IC-701 at £295? Still too much? A Kenwood TS-520S was listed at £225. Want to try QRP for the first time? A Kenwood TS-120V 10W PEP transceiver was listed at £275. Interested in putting on WAB (Worked All Britain) squares on 80m? A Tokyo

Hy-Power HT-180 80m single-band mobile transceiver was offered at just £150. Want a multi-band antenna as well? A Ten-Tec Argosy transceiver with HF5 five-band vertical antenna was for sale at £295.

Don't forget that these transceivers do much the same as brand-new transceivers costing 10 times the amount or more. Maybe you already have an HF transceiver but want to try

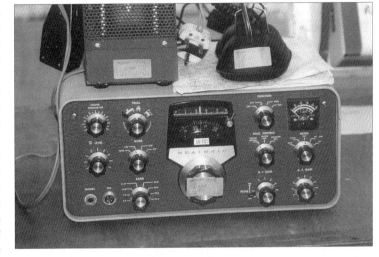

A Heathkit SB-101 HF transceiver is typical of 'seventies equipment which is now available for around £200

some VHF DX work for the first time. Rather than buying a new dedicated VHF multi-mode transceiver, the *RadCom* 'Members' Ads' recently offered two Microwave Modules MMT 28/144 transverters, one at £75 and the other at just £50. A 10m to 4m transverter was also offered for only £50.

If you're interested in 2m or 70cm FM operation, a wide variety of second-hand equipment is available: a Kenwood TR-3200 70cm portable for £130, TH-28E 2m transceiver with 70cm receive capability for £175, or a TR-2200GX 2m FM box for £110. Something for the car? A TM-201A 25W 2m mobile was for sale at £150. Want to try SSB and CW as well as FM? A Yaesu FT-290R was for sale at £215.

But what do all those numbers mean? What is the difference between a TR-2200GX and a TM-201A? Here publications such as *The Rig Review* [1] can be of great use. This lists hundreds of rigs by manufacturer and model number, giving the modes, bands, power output and, where available, year of manufacture and original price. An even better way to find out is to have an expert to give you a bit of help. If you are a member of a radio club, almost certainly there will be a member who will know all the major manufacturers' product numbers for the past 40 years off by heart! Let him know which bands and modes you want to operate, and how much you are prepared to spend, and he should be able to let you know which rigs to look out for.

RALLIES AND JUNK SALES

The 'Members' Ads', useful as they are, are not the only way to pick up second-hand equipment. Radio rallies held around the country throughout the year frequently have a 'bring and buy' stand at which bargains can be found. Go round the rally with an experienced guide so that you don't end up buying something you later regret.

Most radio clubs have an annual 'junk sale' at which real bargains are available. In many cases, prices are rock bottom, with some pieces of equipment going for a nominal sum of £1 or even less. Again, it helps to sit next to

somebody who can advise you on what to look out for – and what to avoid!

Just going to radio club meetings and letting it be known that you are looking out for a cheap transceiver can often bring results. Often, people have pieces of equipment which they almost never use but have never got round to advertising it for sale. Usually, they are only too willing to let it go at a good price to an enthusiastic beginner.

Former PMR equipment like this Pye Europa can often be found at junk sales or rallies. This one was for sale at Longleat in June

Former PMR (private mobile radio) equipment often finds its way to club junk sales. This is equipment from firms which need to use radio to keep in contact with their employees: typically taxi firms, window cleaning companies, security firms, garden contractors and so on. PMR equipment operates on AM or FM on frequencies relatively close to the 4m, 2m and 70cm amateur bands. Some PMR equipment is quite easy to convert to amateur band use and these are often available at junk sales at very low prices. They can be both base stations and mobile units. Again, talk to your radio club expert on such matters for advice on which equipment to look out for and for help in converting them.

DEALERS

A half-way house between buying a brand-new piece of equipment and buying from a junk sale or from 'Members' Ads' is to buy used equipment from a dealer. Generally speaking, prices are somewhat higher than in 'Members' Ads', but the great advantage is that most dealers offer a guarantee, usually for three or six months, on used equipment. Most will also offer a repair service after the guarantee has run out. If you cannot afford the overall price immediately, most dealers are also licensed credit brokers and can offer a number of tempting ways to persuade you to part with your money! A look through the dealers' advertisements in *RadCom* reveals several second-hand HF, VHF and UHF transceivers for under £300. Dealers will usually take your old equipment in part-exchange, so if you have recently gained a Class A licence and want to try HF for the first time you can trade in your old VHF gear to ease the burden of payment.

HOMEBREW AND KITS

Last, but certainly not least, one of the cheapest ways of getting on the air is still to build your own equipment. While the 'Third-method SSB HF transceiver' project (see *RadCom* June and July 1996) is not intended for beginners, simple transmitter circuits have been published in *RadCom's* 'Down to

Earth' column [2] and in *D-i-Y Radio* [3, 4]. The RSGB publication *Practical Transmitters for Novices* [5], while nominally aimed at the Novice licensee, contains ideas suitable for anyone wishing to build simple transmitters. The book includes designs for 160, 80, 6m and microwave transmitters.

But if you don't feel confident enough to build equipment from scratch, a kit may just be the answer. A number of dealers who advertise in *RadCom* sell transceiver kits, and the simple 80m transmitter described by Rev George Dobbs, G3RJV, in reference [3] is based on a kit called a 'Oner'.

CONCLUSION

By now, it should be clear that amateur radio need not be an expensive hobby. Building a rig from scratch can cost as little as a few pounds, a kit a few tens of pounds, or a second-hand transceiver with very similar performance to a new one costing a couple of thousand pounds can be yours for around the £200 mark. Make sure you get guidance from an experienced amateur before you part with your hard-earned cash, especially if you are not absolutely sure of what you are buying.

REFERENCES

[1] *The Rig Review,* Dave Morgan, GW4KYZ. Twrog Press, Penybont, Gellilydan, Blaenau Ffestiniog LL41 4EP. Available from the publisher.
[2] '160m AM transmitter', *RadCom* April 1996.
[3] 'A breadboard 80m CW transmitter', *Electronics Cookbook*, RSGB, p182ff.
[4] 'A 6m transmitter', Ian Keyser, G3ROO, *D-i-Y Radio*, Vol 6, No 3, May-June 1996.
[5] *Practical Transmitters for Novices*, John Case, GW4HWR, RSGB.

QSLing for beginners

They say in certain circles, "The job's not finished until the paperwork is done". That applies to amateur radio too but luckily the sending and receiving of QSL cards can be one of the hobby's great joys.

THE CARD

A QSL card – a typical example is shown in the photo – is a written confirmation that a contact (QSO) took place, sent from one operator to another. It is a kind of final courtesy if you like. Nearly all QSLs take the form of a postcard-sized pre-printed card.

SENDING CARDS

As an RSGB member, you are allowed to send as many QSL cards as you like to anywhere in the world via the RSGB QSL Bureau. Fig 1 shows how the system operates. The only thing it will cost you – apart from the cost of the cards themselves – is the price of the stamp to send them to RSGB HQ. In fact it needn't even cost you that, because QSL cards can be handed in to the RSGB stand at any amateur radio event the Society attends (such as the HF Convention).

For reasons of economy, it's best to wait until you have a number of cards to send, then place them all in one envelope, but *stop*! Here's an important point to remember. To save an enormous amount of time and effort by the

An attractive single-sided QSL card, received as the result of a GB2RS News broadcast

QSL Bureau staff, you need to sort your outgoing cards before sending them. Cards for the UK should be separated from those destined for other countries. Cards for the USA should be sorted by call area, and cards for other countries should be sorted by prefix. Cards for other countries are mailed to the RSGB's sister societies abroad.

RECEIVING CARDS

Although you are bound to receive QSL cards direct through the post on occasion, most QSLs arrive in bundles from 'The Bureau'.

QSL cards from overseas bureaux arrive in Britain at the RSGB HQ. Here, they are pre-sorted and forwarded to the QSL sub-managers. All the

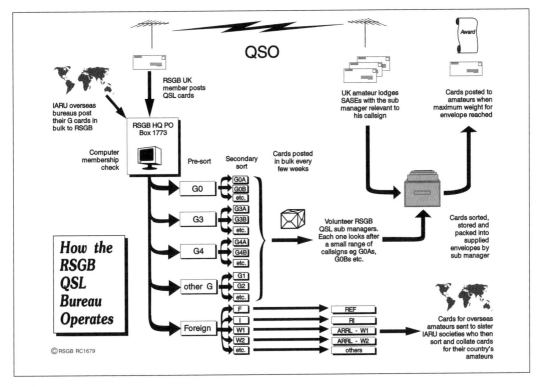

Fig 1. How the RSGB
QSL Bureau oper-
ates

sub-managers – some 90 in all – are volunteers. In order to receive any cards sent to you, you *must* send stamped self-addressed envelopes to your sub-manager. Anyone can *receive* QSLs from the RSGB bureau, but only RSGB members can *send* cards via the RSGB bureau.

People who do not lodge envelopes with their QSL sub-manager are not forwarded any QSL cards. Sub-managers normally hold cards for a while, then dispose of any that appear to be unwanted (they are alleged to make excellent compost). Equally, if you do not wish to receive cards, it is helpful to inform your QSL Sub Manager so that time and space is not wasted looking after any.

Not everyone will respond to QSL cards which you send out. This is an unfortunate fact of life, and one which seems to be exacerbated by your desire to receive a particular card. Sometimes the wheels can be 'oiled' by sending the distant station a stamped addressed envelope direct, an International Reply Coupon (which can be traded-in at any post office in the world for a surface-rate stamp to anywhere else in the world), or a US$1 bill (the so-called 'Green Stamp').

To make life easier, not just for yourself, but for your QSL sub-manager; send several envelopes, each of them numbered (eg '1 of 6', '2 of 6', etc); send your sub-manager envelopes with priced stamps, plus a buffer stock of 1p stamps, which will be used as/when postage rates rise; others send envelopes with class-category stamps (eg 1st, 2nd), so when postage rates rise, the envelope is still properly stamped.

YOUR OWN QSL CARD

When it comes to a QSL card, there are certain items of information which are essential. This applies irrespective of whether you buy blank QSLs on which you have to write all your own details, select one from a swatch provided by a QSL printer and have it customised, or design and print your own from scratch. A QSL card without all of the following might cause the station who receives it confusion or frustration. They are:

- Your callsign.
- The callsign of the station you worked (written on both sides of the card).
- The date and time (UTC/GMT) of the contact.
- The frequency band in use.
- Mode of operation.
- The signal report you sent.
- Your country.
- The magic words "Confirming our QSO", or something similar.

In addition, several optional items of information might be useful. They are:

- Details of your station (transmitter, receiver, antenna, power, etc).
- Your county/region.
- Your WAB area (and book number(s)).
- Your locator.
- The names of any clubs/societies to which you belong.
- ITU zone number.
- Propagation at the time of the contact.
- Comments, or any other relevant information.
- Your name.

Finally, to make your card attractive:

- Artwork (map, cartoon, photograph, etc).
- Use of colour.
- Consider a double-sided card.

These days, plenty of people have the facility to design and print their own QSLs. An A4 sheet of card (coloured?), printed with a modern laser or ink-jet printer, then guillotined as appropriate, can produce QSLs which look really good. The recommended size is 3.5in × 5.5in. Alternatively, look through the adverts at the back of *RadCom* for the commercial producers, then obtain samples.

WHAT ARE QSLS GOOD FOR?

Apart from making the shack look pretty, what can you do with the QSL cards you receive?

Claiming operating awards is the most obvious answer to that question. There are many awards available, some local, some national, some international. Many – but not all – require the operator to substantiate the contacts valid for the award by sending QSLs with the application.

Getting going on top band

Known variously as 'top band', '160m', or the '1.8MHz band', this band has long held a fascination all of its own. Until the recent availability of 73kHz and 136kHz, top band was the lowest-frequency amateur band, its behaviour having more in common with the medium-wave broadcast band than the HF spectrum.

SOME HISTORY

Before the advent of the Class B Licence back in the 'sixties, top band was where most new British amateurs cut their teeth, often with amplitude modulation (AM), and using the band for local nets. Just a few hardy (or perhaps foolhardy) folks considered top band a DX band, offering the opportunity to work stations in other countries or continents. However, until the late 'seventies and even the 'eighties, many countries had no 160m allocation, mainly because this part of the spectrum was still used primarily for maritime purposes. Belgian amateurs, for example, were only given a 160m allocation as recently as 1 January 1987.

Even in the UK, the band was shared with maritime stations (ship-to-shore and Loran navigation beacons), and amateurs were limited to 10W input on AM and CW, and 26W output on SSB. These power levels were more than adequate for chatting around the country but made DX-working a serious challenge, especially as what might be considered a 'full-size' antenna, such as a half-wave dipole or a quarter-wave vertical, was (and remains) beyond the scope of most operators. A half-wave dipole is about 260ft long, and a quarter-wave vertical some 130ft high!

Of course, to some operators, these limitations made top band exactly the sort of challenge they were looking for, and there were regular 'trans-Atlantic tests' organised by the late Stew Perry, W1BB, doyen of top band operators, well into the 'seventies, in which the better-equipped 160m operators would try to help newcomers to make their first trans-Atlantic contact on the band.

Going perhaps to the other extreme, top band used to be a popular band for mobile operation, and it was relatively common to see amateurs with 8 or 10ft loaded-whip antennas attached to their car bumpers!

CURRENT STATUS

Nowadays, the Loran beacons have closed down, having been replaced by more modern systems, especially GPS, the satellite global positioning system. This has freed up spectrum on 160m, and every country now allows its amateurs access to the band, although the exact frequency allocation and power limits vary considerably from country to country. Here in the UK, the power limit in our primary allocation is 26dBW (400W), but only 15dBW (36W) in the shared part of the band. All the usual modes of transmission are permitted. Most overseas allocations overlap with the UK primary allocation, so this part of the band tends to be used for international working, with Region 1 of the IARU recommending that SSB activity should not take place below 1840kHz. The higher part of the band (above 1900kHz) is more typically used for intra-UK activity such as the Worked All Britain group nets, GB2RS news broadcasts, and local 'ragchews' (but in the winter, do try to avoid causing interference to the limited Japanese allocation, see below).

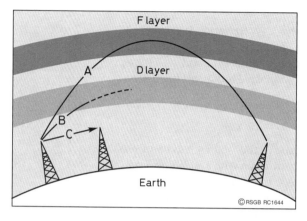

PROPAGATION

As mentioned earlier, top band propagation has much in common with the medium-wave broadcast band, being limited in range during the day but extending out to hundreds, or even thousands, of miles during the hours of darkness. This is because top band signals are absorbed by the D layer of the ionosphere (Fig 1), which exists during daylight but disappears at night. After dark, signals are reflected from the F layer, and can travel long distances.

Fig 1. At night, top band signals 'bounce' off the F-layer of the ionosphere for DX contacts (A). During the day, the Sun's rays ionise the D-layer which absorbs sky waves (B), so most day-time contacts are by ground wave (C)

What this means is that top band is ideal during the day for cross-town contacts, with very little interference from more distant stations. Ground-wave absorption is less than on the higher-frequency bands, and reliable daytime ground-wave contacts can be made out to 50 miles or so. At night, you should be able to make contacts around the country, and to the nearer parts of mainland Europe, even with modest power levels and limited antennas. If you fancy doing some serious DXing on top band, this requires a bit more effort. As with long-distance propagation on the higher-frequency bands, DX working on top band requires launching signals from your antenna at a low angle of elevation. To do this with a horizontal antenna *ideally* needs the antenna to be at least a half-wavelength above the ground – relatively easy on the high bands (30ft or so on 20m, for example) but quite unrealistic on top band where a half-wavelength is 260ft or so! The alternative is to put up some sort of vertical antenna (see below) which will have a much lower angle of radiation than a low dipole. A vertical antenna will also be very suitable for daytime ground-wave contacts, as described above.

However, there is another trick which top band DX enthusiasts learn early on. This is that, at the boundary between daylight and darkness, there is a 'tilting' of the ionosphere, because the various ionised layers rise in daylight and are lower at night. This tilting can be used to advantage, with high-angle signals from your antenna being reflected onwards to the distant end at a much lower angle to the horizon.

A related effect is that of so-called *greyline* propagation, where signals propagate along the daylight/darkness boundary (the *twilight zone*), with very low attenuation.

Long-distance 'openings' on top band can be very short indeed, with signals from, say, Western Australia coming up out of the noise, peaking at a good level at Australian sunrise, and disappearing back into the noise, all within the space of five to 10 minutes.

And finally, top band signals are absorbed more readily than higher frequencies in the auroral zone. In consequence, paths crossing the auroral zone (such as from the UK to Alaska and Western Canada) are very difficult, and quite impossible when the A and K indices are high. What all this means is that, if you want to use top band for local and semi-local contacts, you need not worry too much about the subtleties of propagation. The band will be very reliable for ground-wave propagation during the day and for high-angle sky-wave propagation at night. For long-distance working, you will need to pay attention not so much to the solar flux, as on the higher bands, but more to the A and K indices and to sunrise and sunset times which can, for example, be calculated by computer programs such as Geoclock or read directly off a propagation aid such as the DX Edge [1].

EQUIPMENT AND ANTENNAS

For local contacts on top band, signal strengths will usually be good, and relatively simple equipment will suffice. This makes the band very suitable for homebrew equipment and, in days gone by, every amateur radio magazine used to feature designs for simple top band transmitters and receivers. Nowadays, all HF transceivers cover top band but some experimenters still prefer to build their own equipment, even continuing to use amplitude modulation, or double-sideband reduced carrier (DSB). Receiver sensitivity is not usually an issue, as band noise is higher than on the high bands.

However, if you are going to get involved in DX-chasing, other equipment parameters start to become important. Perhaps the main one is susceptibility to intermodulation, as local signals can be very strong indeed, whereas DX signals on an adjacent frequency can be hovering around the noise level. Top band DXers who have the space available deal with this to some extent by using specialist receiving antennas such as the Beverage (see below), and this places another requirement on their equipment – the ability to use separate receive and transmit antennas.

Another way to deal with the vast range of signal strengths encountered on the band is to operate split frequency. In the past, this meant that the band was, by general consent, divided into so-called DX 'windows', with European stations transmitting in one part of the band, North American stations in another, and so on. Nowadays, split-frequency operation is more likely to mean a DX station listening a few kilohertz away from his transmit frequency in order that he can be heard clearly without interference from others calling him. This means you will need a radio capable of split-frequency operation. The extreme instance of this is when chasing Japanese contacts on top band, as Japanese stations have a 160m allocation of 1907.5 to 1912.5kHz but usually listen around 1830kHz for European callers.

If you are pressed for antenna space, don't worry. There are many ways to put out a top band signal sufficient to work around the UK and Europe. For close-in working, a low dipole is more than adequate, and you can easily shorten the overall length by adding some inductive loading. If you already have a trapped dipole or G5RV, one solution is to place parallel resonant traps for 80m at the ends of your existing antenna, and then extend the antenna at the ends. The 80m traps will act as inductive loading on top band.

Alternatively, and especially useful if you want to generate some lower-angle signals for longer-distance working, the popular solution is to 'strap' the feeder of your high-bands dipole (in other words, to short the inner and outer of the coax together), and then to feed this as a shortened-vertical against a suitable earth system. The dipole itself will act as capacitive top-loading, which will have the effect of electrically lengthening the vertical part, making it more efficient. As with any vertical antenna, the earth system requires more than a simple electrical earth. Run out as many earth radials as you can manage, as long as possible. Many top band enthusiasts have developed ingenious means for burying these radials in the lawn with the minimum of effort.

Another handy solution to achieve a vertical radiator, if you happen to have a metal mast supporting an HF or VHF beam, is to shunt-feed the mast, again against a good earth system. The beam will act as capacitive top-loading. Shunt feeding is covered extensively in the literature, for example in ON4UN's popular book on low-band DXing [2].

Because distant signals on top band can be very weak, and gain antennas beyond the scope of most amateurs, a number of alternative solutions have been used for improving reception. The Beverage antenna, named after its inventor, is a very long wire (two or three wavelengths), close to the ground and terminated at the far end with a non-inductive resistor. Beverage antennas are inefficient, and therefore unsuitable for transmitting purposes, but discriminate against high-angle signals in favour of those received at low angles, which makes them ideal for DX receiving. However, their length precludes their use for most of us (unless you have a disused railway line at the bottom of your garden, as one top band enthusiast used to have!). Alternative receiving antennas include tuned receiving loops, which can be just a few feet in diameter [3], and EWE antennas, which were written up in QST magazine [4]. If you don't have space for any of these, an old trick on top band, if you are having trouble hearing the distant station, is to try receiving on any other antenna you have, such as a high-bands dipole. Sometimes you will find that this does the trick nicely.

CONTEST AND DX WORKING

If you develop the top band 'bug', you may well want to join in some of the regular operating events which take place on the band. These include the RSGB Club Calls Contest, the RSGB LF Cumulative Contest, and the ever-popular RSGB 1.8MHz CW Contests, which take place in February and November. Internationally, the most popular top band contest is the CQ World-Wide 160m CW Contest, which runs over the last full weekend of January, and the SSB leg of the same event, which usually runs over the last full weekend of February .

And if you start to chase 160m DX, then there are several goals to which you can aspire, including top band versions of several of the popular awards such as DXCC, Worked All Zones and Worked All States. Most major DXpeditions nowadays make sure they are on top band with good signals, and you will be surprised what you can work. 1997's VK0IR Heard Island expeditioners, for example, managed over 1200 QSOs on top band, despite being over 4000 miles from any major population centre, and this included 39 contacts with the UK. Mind you, the group did use four phased vertical antennas, with the added advantage of being located right next to salt water.

It is also worth noting that top band is a year-round band, even though the top band 'season' is often regarded as running from September to April. Our summer months coincide with winter conditions in the southern hemisphere, and some of the best propagation into southern Africa, South America and even Australia can occur at the height of our summer. The only snag is that noise levels on the band are likely to be higher, due to electrical storms, but the serious top band enthusiast is ever-patient, checking the band night after night for that occasion when noise levels are low and the DX is there for the taking.

REFERENCES

[1] The DX Edge, produced by Xantek, is available from the RSGB Shop.
[2] *Low Band DXing*, 2nd edn, ON4UN, from the RSGB Shop.
[3] 'A receiving antenna that rejects local noise', K6STI, *QST* September 1995.
[4] 'Is the EWE for you?', WA2WVL, *QST* February 1995.

FURTHER READING

The ARRL Antenna Book, 18th edn.
Amateur Radio Operating Manual, 4th edn, ed G4FTJ.
Both are available from the RSGB Shop.

A guide to HF contesting

You don't have to be a Novice or newly licensed amateur to want to find out about HF contests – the bug can bite at any time, even after many years of being on the air. Here you will find out how to get started for the first time and how to improve your operating skills.

COMPROMISE NECESSARY

No matter how recently your interest in amateur radio has developed, you cannot have failed to come across a contest when tuning the HF bands. There are few weekends in the year when there is no contest activity on SSB, CW, PSK31 or RTTY. Generally speaking, because of the nature of propagation, HF contests tend to be busier than VHF contests, but the same basic skills are involved and an operator who is successful in one is likely to do well in the other.

The amount of activity generated can, and does, cause controversy. Some people complain that the presence of a contest spoils their enjoyment of the hobby, as they are unable to find space to have their usual 'ragchew'. However, the very large amount of activity generated is proof of the popularity of contests and those wishing to take part feel they are entitled to the recognition of their aspect of the hobby. This is not to say that the needs of non-contesters should be disregarded but, love them or hate them, contests have been established for more than 60 years and are here to stay. The International Amateur Radio Union (IARU) decided that the WARC bands (10, 18 and 24MHz) shall be totally free of contests in order that non-contesters have completely clear bands in which to operate. Additionally, individual member societies of IARU (such as the RSGB) have established their own rules which stipulate frequency limits in order that a portion of each band is left clear of contest traffic. Probably the only remaining exceptions to this practice are the major contests, such as *CQ* World Wide, which attract truly worldwide participation and place great demand on all available space.

WHY ENTER CONTESTS?

The reasons for participation in contests are many and varied. The main reasons are to satisfy the basic competitive instinct which is inherent in the majority of us, to see how your station performs against those of your friends or rivals, the thrill of seeing your callsign in print, and in future years watching your call climbing up the results table. You will come to look forward each year to your own favourite contest, one that suits your station or interests. Whatever level you attain in contesting it will most certainly make you a more confident, proficient and skilled operator.

Don't forget that you can *participate* in a contest without *entering* it. There will be stations on the air in a contest which you will never see at any other time. What better way to enhance your DX totals?

Not everyone can be a winner. For one reason or other it is not easy to get a level playing field – there will always be someone who has a bigger or higher antenna, greater power, better location, or better propagation. Striving to make the most of what you have is what brings the greatest satisfaction. There are always goals and targets to be set and achieved, such as beating last year's score overall and on each individual band.

In the case of RSGB contests, an equipment code has been devised which accompanies each listing in the results table. This will show you what you are up against and enable you to compare your performance against similarly equipped stations. This code gives the height and type of antenna in use, the number of elements and the output power used. Details of this code can be found in the 'RSGB Contesting Guide' which is published in the January *RadCom* each year.

The RSGB HF Contests Committee (HFCC) has introduced restricted sections in many of our events, with the purpose of increasing the activity and competition amongst stations with the 'average' set-up, ie those who reside in built-up areas or have small plots. These restrictions relate to power output and antenna height. There are also categories for low-power (QRP) operators. Although we only produce one listing for the results, it can be quite readily seen from the results table and equipment code the category of each station, and certificates are awarded to the leading stations in each section.

The Bristol Contest Group operating as G6YB/P in NFD and on 50MHz in June 1996. Antennas are 4-element beams for 21 and 14MHz, 7-element 50MHz beam, 4-element 28MHz and 2-element 7MHz beams

HOW DO I START?

With the exception of National Field Day, there is no necessity to register in advance for any HF contest that you may come across. Your participation will be most welcome and indeed desired by those you hear calling, so go right ahead and join in.

The first important thing to do is to *listen* carefully to a few contacts. Establish, if you can, what the event is and what 'exchange' is needed from you. You will need to know the basic rules and (except for the smallest of national contests) this information can be found in the current issue of *RadCom*. First, refer to the Tim Kirby, G4VXE, 'Contest' column. Each month he publishes a 'Contest Calendar' for the current month, and the odds are that the contest you have heard will be mentioned there. If it is an RSGB event you can then refer to the 'RSGB Contesting Guide' for the full rules. If it is an overseas contest, reference should be made to the 'HF' column of Don Field, G3XTT, which includes brief rules of foreign contests for that month.

Most 'exchanges' require a report, eg 59 or 599. Now that computer logging has become the norm and where the 59 is already in the 'report' field, just waiting to be entered, you are not going to be too popular with a station who is working three or four contacts per minute if you give a 46 or 478 report; it is better to accept that this part of the exchange has become meaningless, but is still necessary! QRP and some other contests still require accurate reports, which is just as it should be. In addition you will, in most cases, need to include a progressive serial number, commencing at 001. If it is an RSGB contest, you may also be required to give your 'county code', which again can be found in the 'RSGB Contesting Guide'.

Instead of a serial number, you may be required to give your power, your CQ or ITU zone, your IOTA reference, or even your age! With only a little experience you will soon get the hang of it and if you get it wrong – no problem – the other station will tell you what is required of you.

Most operators dip a toe in the water by tuning through the band answering CQ calls. There is an undeniable thrill in raising distant stations or getting through against a number of other callers. However, unless the event is a big one, you are likely to spend an increasing amount of time searching for new stations to work, and the time then comes when you decide to get really involved by finding a clear frequency and make your first 'CQ contest' calls.

Unless you have a very-well-equipped station, don't expect miracles! However, there will come a time when you, as a British station, will be in demand and two or three callers, perhaps more, will come back at the same time. Now you're on the hook – your first mini pile-up. Don't panic!

CONSIDERATION AND COMPETENCE

On the HF bands you may have heard an operator working from a rare country. Everyone wants to work him but very few are making it, so the frustrations of a few boil over into anger and we get the bad behaviour of those operators who make it impossible for anybody to get a contact. A good operator keeps his temper. Competent handling keeps waiting stations in anticipation, in the belief that they have a chance of a contact. Consequently, they tend not to misbehave. The ability to handle a pile-up seldom comes as a natural ability but is gained by practice and experience. There is no substitute for listening. The next time you come across a pile-up, note how the operator deals with it, analyse your reaction to the style; are you spell-bound in admiration or are you critical?

Back to your contest now. Resolve not to be like that DX operator, listen carefully and try to pull something useful out of the pile-up. Often one caller is slightly behind the rest or just one word will often do. Saying something like "the station with Juliet in the call, you're 59 040" will isolate him, get you the contact, and, followed by a simple "QRZ?" bring in the other stations who will have waited. Uncertain handling at the outset, asking for repeats, would have sent them on their way up the band looking for more efficient stations to work.

CW CONTESTS

CW contesting is perhaps more daunting for the beginner, mainly due to the sending speed. The keyers employed in the logging programs are usually set somewhere between 28 and 32WPM. This can be a real demotivator, but do not be deterred.

For the Novice licensee, or those newly on the air with a full A or A/B licence, the RSGB has designed a special series of contests for operators who would like to try out contesting in a series of relaxed, slow-speed events. Known as the 'Slow-Speed Cumulatives', these events are held in the spring and the autumn and comprise five sessions in each event. The duration of each occasion is 90 minutes and there is a maximum allowable speed of 12WPM.

The sessions are arranged over a five-week period and staggered one day each week, commencing on Monday the first week, Tuesday the next week, and so on. This way each day Monday to Friday is covered and, as this is likely to cover the night of your local club meeting, we should like to encourage participation from club stations, where guidance from experienced operators will also be on hand. You can enter from the club either as a single operator or use more than one operator in the 'multi-op' category. If you don't have your own station or club facilities, consider asking another amateur if you may use his station. Read the rules in the 'Contesting Guide' carefully and note the relevant power limits stated.

Having mastered the basic format, you will be ready for greater things. Speed comes with practice, so just remember that all the top-flight operators that you will come across started exactly where you are now. The other operator does not want to know your name or QTH, nor is this the time to ask for QSL information. He isn't being impolite but he is in a contest and so time is of the essence.

If you choose to try a few CQ calls, on no account send faster than you can comfortably receive. This may sound obvious but can easily happen if using keyboard or memory keyer sending, and the machine gun that will surely hit you can spread an awful lot of egg over your face. The last thing you need at this stage is damage to your confidence. When sending at your own speed, you will find that calling stations will more or less match your speed. Don't be afraid to ask them to "QRS PSE" if they are too fast; it's your frequency and you call the shots.

RSGB CONTESTS

Contests can be divided broadly into international and national events. International contests such as *CQ* World Wide are usually 48-hour events where everyone works everyone. The RSGB takes a different approach. Being of the opinion that there already sufficient all-band contests, it has gone the route of promoting single- or dual-band contests, with activity restricted to 12 or 18 hours duration. There are, of course, exceptions. First, the Commonwealth Contest is on all bands, 80–10m, and lasts for 24 hours. As participation by definition is very limited, contact rates are very slow and a lot of skill is required to obtain a good result, and little disruption is caused to non-participants. The other exception is the RSGB IOTA Contest, now just four years old. This is a 24-hour all-band worldwide affair; it does not specifically promote UK stations but puts the emphasis on everyone contacting islands of the world on an equal basis. Different in format and concept from other contests this event has been warmly received in all areas of the world and is already the RSGB 'flagship' event with an entry rate growing very rapidly.

On the home front, the RSGB HF Contests Committee has been keen to foster the spirit and activity of contesting among the membership and to try to offer something to all of the different interest groups. Take a careful look at the 'RSGB Contesting Guide' and note the preamble to each of the events described there. The slow-speed contests have already been mentioned, but there is a special slow-speed segment of the band in the Affiliated Societies Contest, and the LF Cumulatives have been designed as training periods for stations wishing to progress to higher profile contesting. It is in these events that many of us started our involvement with contesting.

For SSB operators, the Club Calls Contest is primarily for club stations, although individual entrants are also welcomed, and we would like to see more clubs taking part in the 21/28MHz SSB Contest. In fact it will be noted that multi-operator categories have been added to most of our contests to encourage participation from clubs, the newly licensed, and others who may not have their own equipment.

The contests mentioned in the previous few paragraphs are all mainly low-key events, designed to get you started. Many of you reading this now are new to the hobby and are our future, our lifeblood. All the successful operators that you hear on the bands began just like you, wondering what it was all about. The UK may not rank among the top nations in the world for quantity of contest operators, but we are right up there when it comes to *quality* and we wish to continue that tradition. We would like to increase the *quantity* too, for it is a fact that many countries with far smaller populations, such as Finland and the former Yugoslavian states, leave us far behind when it comes to numbers participating in the major contests. If your club takes part in NFD or SSB Field Day, *do* make an effort to be part of it. You will be welcome, and learning first hand from experienced operators is by far the best way to learn the ropes. If the club does not currently take part, why not organise something yourself from amongst the membership? The odds are that there will be at least some antenna equipment belonging to the club tucked away somewhere. You may not finish top of the table, but you will learn a lot, you will have a lot of fun and memories that will last a lifetime, and that is as important as winning.

The purpose of the RSGB HF Contests Committee (HFCC) is not only to decide policy and arrange the contest programme on behalf of the Society, but also to meet the needs and desires of the membership. Your views and feedback on what the Committee does are welcome; you are invited to contact members of the Committee for advice or guidance.

INTERPRETING CONTEST REPORTS

We try to publish useful hints and tips in the contest reports where space allows, and there are always detailed reports of NFD, SSB Field Day and the IOTA Contest. Then there is the *RadCom* 'Contest' column, already mentioned, which contains current happenings, trends and developments, as well as suggestions on how to improve your station.

When reading contest reports, such as those in the 'Contest' column, frequent reference will be found to tactics adopted by the operators. Some of the terms used may require explanation:

Run frequency

Finding a clear frequency, calling CQ and working a stream of callers is known as *running* and the frequency used is the *run frequency*. Unless you have a commanding signal, it will soon be found that success with this method is short-lived, for someone with a better signal will park right on top of you and he may not hear you (or, worse still, he may not wish to hear you). Whatever the reason, if he doesn't move after the first time of asking, you may as well forget it and find another spot. Good operators will move but it may be that he genuinely does not hear you. Time spent arguing the toss is completely wasted, and likely to put you in a bad mood, serving only to delay your progress in the contest. Accept the fact that this is going to happen (it happens to the 'big guns' at times so it's bound to happen to you), so the sooner you move away and start afresh the better.

When you have a good antenna or enjoy the peak of propagation on a band that is open to all areas, a good 'run' can be an exhilarating experience, limited only by the speed at which you can work callers. It is common for DXpedition-type contest stations to work in excess of 200 stations per hour.

Searching and pouncing (S&P)

This involves combing the band and checking each station heard either for a new contact or for a new multiplier. This tactic needs to be methodical and accomplished as quickly as possible. It is very easy to spend too much time calling a needed multiplier to the detriment of your overall performance, so it is necessary to have a plan of campaign.

Keep a note of stations that you don't raise after a few calls. If your rig has memories it is an advantage to store them, then you can return again at will and be bang on the frequency. So how long should you call? You need to make a judgment based upon the rules of the contest. For example, let us take two contests run by the American magazine *CQ*, probably the two busiest contests of the year.

The first is *CQ World Wide*. Suppose the desired station is SU1ER in Egypt. First, you can be pretty sure that you are unlikely to run across another Egyptian station, and it is a much-needed DXCC multiplier. In addition, SU is in *CQ* Zone 34, and in this contest zones *also* count for multipliers, so he is worth two multipliers. Experience will soon tell you that active stations from Zone 34 are also very thin on the ground.

If, however, the contest is *CQ WPX*, he would not be so important. In that contest, only the prefix counts as a multiplier – and then only once and not on each band. So a simple DJ5 prefix is equally important if you don't have one, and your time would be more productively spent looking for other new prefixes which are easier to work.

'Mults'

The importance of multipliers cannot be over-emphasised. Obtaining the maximum possible number is essential to obtain a good score. Often it is not the station with the greatest number of contacts who is the winner but the station with the most multipliers.

An illustration will make this clear: station A has 300 QSOs and 30 multipliers, station B has 250 QSOs and 37 multipliers. Awarding three points per QSO we find that A has a final score of $300 \times 3 \times 30 = 27,000$ points, whereas B has $250 \times 3 \times 37 = 27,750$ points and claims the prize. It is therefore a matter of striking the right balance between running and S&P.

When listening to experienced operators you will notice that they don't miss an opportunity to make new multipliers and you may have heard them referring to 'moving people around'. When a rare multiplier is running a frequency, some of the callers will be heard to ask "can you QSY to 15m?". If the caller himself is rare DX there may be a positive response and you are left with an empty frequency, cursing a missed opportunity and wondering if you will find him again on that band. Your reaction should be to follow the QSY – after all, you have noted the new frequency and that is hot information. You can make it before he has time to build a new pile-up and even before the Packet Cluster followers get to him.

If the DX station calls *you*, then you must think on your feet and grasp the opportunity. Assess which other bands are likely to have a path at that time and ask him to QSY to the greatest probability band first, followed by other needed bands. If you can do this with all needed multipliers for any country or area with relatively low amateur radio activity, it will pay dividends because you may never run across them again in the contest. Desired stations can be very hard to locate on a busy band.

'Skeds'

Scheduled contacts ('skeds') made during the contest, are common practice. However, skeds made *prior* to the event would be a breach of the spirit (and sometimes the rules) of the contest. If you contact a VK on, say, 20m during the daytime, it may not be possible to move him to any other band instantly. However, you do know that there is a good probability of making a contact with that area at 1730UTC on 7MHz, so you make a sked with him and even though the going may be tough at the time of the sked, the fact that you are listening for each other can result in a QSO and multiplier that may not otherwise have been obtained.

Skeds should also be tried with other hard-to-find multiplier stations for the higher or lower bands as the case may be. Modern computer logging programs will display a warning message on the screen one minute before the sked time.

AFTER THE CONTEST

If your approach to the contest was casual, you can simply close the log on it after the event – you are under no obligation to submit an entry. However, if you made more than a dozen or so contacts, the adjudicators would appreciate a check log from you, because the more information they are given, the more accurately they can score the logs submitted.

But, if you have decided that you *do* wish to submit an entry, what do you do next? First, it will be necessary to compile the log to be submitted. Here the 'RSGB Contesting Guide' gives comprehensive details of the format required. Assuming that your first efforts are on paper, printed examples of log sheets can be found in recent editions of the *RSGB Yearbook.*

Scoring the log

Reference to the contest rules will give you the basis upon which to score your log. Each contact is allocated a basic number of points. Imagine that you have had 50 contacts spread equally over five bands with a basic points-per-contact of 5. First of all, check the contacts for each band. If you have contacted a station more than once on the same band, points should not be claimed for the subsequent contact and it should be marked "Dupe", meaning duplicate. The penalty for failure to observe this rule can be severe; RSGB policy is to deduct 10 times the number of points claimed for that contact and in this example that penalty would be equal to the entire number of points claimed for that band! The next stage is to check the multipliers or bonuses.

Mark each different multiplier. For instance, if your 10 contacts on one band include four from USA, two from Canada plus four other countries, you will have a total of six multipliers for that band. Total the points for each band, then add them all together, then do the same for the multipliers. In our example you will have 50 points for each band × 5 = 250 points for total points score. If your multiplier total is 40, you would have a final score of 10,000 points, ie 250 × 40.

If bonuses, rather than multipliers, were used the bonus contacts would be totalled and multiplied by their value. So if the bonus was worth 10 points per contact the final score would be 250 + (40 × 10) or 650 points.

SUBMITTING THE ENTRY

All that remains to be done is to transfer this information to a summary sheet, an example of which can also be found in the *RSGB Yearbook.* This sheet, together with the log sheets, a dupe sheet, plus a list of the multipliers being claimed, can now be posted off to the organisers. Don't send it by recorded or registered mail. If you require confirmation of safe receipt enclose an SAE or postcard with a stamp or International Reply Coupon (IRC).

The results of RSGB contests are published in *RadCom*; if it is an overseas contest, the scores of British entrants are often given by G3XTT in the 'HF' column. If you want to received detailed results from overseas contests, you should enclose an A4 or A5 SAE with one or two IRCs.

Duplicates

Some contests require you to submit a dupe sheet when you have made, say, 200 contacts, or more than 50 contacts on any one band. This is good practice anyway, and helpful to the adjudicator. What is required is a separate list of contacts for each band arranged in alphanumeric order with your contact number shown for each call. These lists immediately highlight the duplicated contacts.

As your experience increases it will be found that it is not difficult to make 100-plus contacts on a single band in one of the busier contests, and that compiling dupe sheets is a time-consuming, and – if several hundred callsigns are involved – a very tedious business.

Maintaining a dupe sheet as you go along is the answer to this problem when logging on paper. It requires a large sheet of paper, say A3, for each band, ruled into 26 box sections, one for each letter of the alphabet. The calls logged are recorded on the dupe sheet by suffix. For example, G2ABC would be entered under 'A' and HG65N would be entered under 'N'.

This method more or less spreads the information evenly over the sheet, making it quicker to find a given station. Listing under prefix would result in a long list of stations under letters such as D, G or W, and you would quickly run out of space in those sections, whilst the other boxes remained almost empty.

Handling the dupe sheet as a single operator, unless it is a very slow contest, is always a difficult task, seldom fully managed. If enjoying a good run, with plenty of callers coming at you, do you work them at three or more per minute or slow it down to half that rate while you attend to the dupe sheet? You must also keep a multiplier sheet up to date at all costs. I must confess in my own case the dupe sheet is put aside, and brought up to date as best I can during slow periods. They are seldom completed satisfactorily, for a good run can soon put a couple of hundred contacts in the log, which takes some catching up on, and is also not the best use of time when there are more stations to be worked and other bands to be checked.

These shortcomings are the reason why the large majority of contesters now use computer logging.

COMPUTER LOGGING

It is not many years since we were all logging on paper. In an all-band contest we required log sheets, dupe sheets and multiplier lists for each band – so much paper that it took up more room than the equipment! The amount of paperwork after a contest finished did in fact deter many people from submitting an entry. A real breakthrough came with the advent of real-time computer logging, which gives instantaneous warning of duplicate contacts and even scores the entry for you.

The logging programs are now so sophisticated it is a real joy to contest using these tools. The score is computed as the contest progresses, giving an up-to-the minute picture, and there are windows which detail which mults have been worked and which are still needed on each band.

There are many programs available, ranging from the very simple to the ultimate. The latter type includes *CT* by K1EA, which has everything including computer control of the rig, support for the Packet Cluster, networking facilities for multi-op stations and even a 'band map' of the band in use which details stations worked and highlights packet spots still needed.

Probably the most suitable program for participants in RSGB contests is *SD* (Super Duper) by Paul O'Kane, EI5DI, because all RSGB contests are supported. It is very user-friendly too – important if your keyboard skills are not highly developed.

ADJUDICATION

After submission, the logs are checked by the contest adjudicator. His task is to compile a database of all the callsigns appearing in the submitted logs and then check each individual log for accuracy. Special attention is given to any call which appears in one log but not in any others. Such calls may be miscopied in which case they are known as *broken* or *busted* calls, and all claimed points are deducted for that contact.

If that contact happened to also be a multiplier, another contact with the same multiplier will be looked for, so it is a good idea to include 'spare' multipliers on the multiplier list for this purpose. If the call differs greatly and cannot be regarded as broken, then it is referred to as a *unique call*. Your log will be cross-checked against others, and it often happens that your call may not appear in the other log. Even though you *think* you worked him, he was in fact contacting someone else whom you could not hear. This type of error will decrease with experience but can still occur. Even top DXers get some "not in log" responses to QSL card requests. Errors in copying any of exchange data is penalised by the deduction of one third of the QSO points per error.

Mention has been made of the punitive penalties applied in the case of unmarked dupes but an excessive number of unremoved dupes or unverifiable contacts can lead to disqualification. It is clear that entries need to be both clear and accurate.

Many errors occur during the transfer from original sheets to entry sheets, due to misreading handwriting or careless typing. In the case of RSGB contests, the adjudicators often comment on their amazement at how some operators will put in 24 hours or more hard work on the contest, but not even glance over the entry for errors after preparation before submitting it.

FINALE

It has been said that more can be learned about propagation during a full weekend contest than in a full 12 months of casual operating. You will discover the openings on the separate bands to the various areas of the world which occur at different times on a range of frequencies. Knowing when and where to look will give you an edge over many of your fellow operators when it comes to looking for DX in casual operating sessions or chasing that rare station in pile-ups.

The improvements to your station as you strive to improve your position and technique will come along gradually as you change, hone and fine-tune the antennas, equipment and layout of the station.

Most of all, you should obtain enjoyment and satisfaction from contesting. You can get a real buzz from a good event because, above all, contesting is about having fun from your hobby.

Your first use of a repeater

This comprehensive article explains why repeaters are necessary for good, consistent contacts on VHF and UHF from moving vehicles, operation which we know by the term 'mobile' or '/M'.

Not many radio articles start with "Once upon a time" but this one does!

"Once upon a time, in two far-off valleys, there lived two tribes, separated by a mountain. On this mountain lived a great giant, who forbade any contact between the two peoples unless it was by his permission. For a member of one tribe to speak with a member of the other, they had to call upon the giant to relay their message.

"Now the giant was a sleepy giant, and often did not hear members of the two tribes calling him. To overcome this difficulty, members of each tribe decided to blow upon a whistle to alert the giant, who would then wake up and repeat their messages across the great mountain.

"Some tribesmen, being impatient to wait their turn from the giant (because he would only take one message at a time), tried to make contact directly. However, such was the distance between the two valleys, that only the tribesmen with the loudest voices could make themselves heard, and then only by climbing as high above the valley floor as possible.

"So the people of the two valleys found their giant of such great help, that, as the days went by, other giants were invited to come to the great mountain and pass messages to and fro."

This little tale shows what a repeater does. The giant is the repeater, and the tribesmen are individual amateurs, mobile, driving around. And yes, repeaters are often located, if not on a mountain, then certainly on high ground.

WHAT IS A REPEATER?

A repeater is a device which will receive a signal on one frequency, and simultaneously transmit it on another frequency. Careful design has meant that repeaters can receive and transmit within the same band. This means that the same antenna can be used for both reception and transmission.

In effect, the receiving and transmitting coverage of the mobile station becomes that of the repeater and, since the repeater is favourably sited on high ground or a tall mast (Fig 1), the range is greatly improved over that of unassisted, or 'simplex', operation (Fig 2). The coverage areas of two mobile stations is always changing shape, whereas the coverage of a repeater will stay constant, and can even be published.

HISTORICALLY SPEAKING

It is in the early 'seventies that the history of amateur repeaters begins in the UK. The influx of compact, Japanese-origin transceivers operating on narrow-band FM allowed amateurs for the first time to fit such units easily in

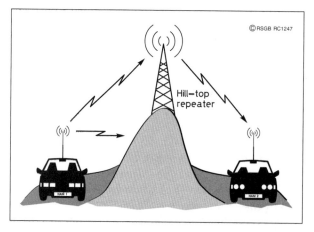

©RSGB RC1247

Hill−top
repeater

HAM 1

HAM 2

Fig 1. Because of the terrain, the two mobiles would not be able to achieve communication from one to the other without the assistance of the hill-top repeater

Fig 2. The simplex coverage areas of mobile stations A and B are constantly changing as the two vehicles pass through different terrain. The repeater allows communication between A and B anywhere within the repeater's coverage area

their cars. Previously, more cumbersome valve-operated sets had been used, but were hardly conducive to easy communications.

With some experience gained from the private mobile radio (PMR) business, it was soon realised how a strategically sited repeater could allow relatively low-powered sets in cars to communicate over relatively long distances. The first repeater in England was GB3PI, located near Cambridge, which was soon followed by others on both 2m and 70cm.

Realising that there was potential for a nationwide network of repeaters, the RSGB created the forerunner of the Repeater Management Group (RMG) to oversee the specifications, planning, and vetting of applications.

The actual building of the repeaters was undertaken by local groups of amateurs, and a whole new 'club' infrastructure was created of repeater groups recruiting members for an annual subscription. The groups would build, maintain and operate repeaters in their area on behalf of their members. Although use of the repeaters was open to all licensed amateurs, those who used the repeaters would be morally expected to join their local group, or to subscribe in some way. Some groups were very well organised, publishing professional newsletters, holding regular meetings, and making many technical developments on their repeater systems.

At the core of each repeater group was the technical team who would have access to the necessary equipment to keep these repeaters operating within the required specifications.

TECHNICAL CONSTRAINTS

Technically, a repeater consists of a receiver, a transmitter, antennas, filtering, and control logic (Fig 3). Each repeater is allocated a 'channel', consisting of a receive frequency and a transmit frequency within the same band, but separated by 600kHz (on 2m) or 1.6MHz (on 70cm). A 1750Hz toneburst on the input frequency was employed to 'wake up' the giant – sorry, repeater – when it would then relay the input audio to the transmitter. The identifying callsign (of the format GB3xx) of the repeater is transmitted in Morse, and if no audio is detected after, say, 10 seconds, the transmitter carrier shuts down.

The major technical constraint to the repeater builders over the years has been

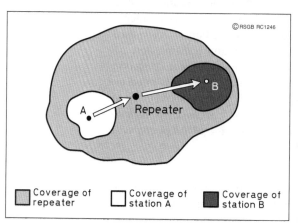

©RSGB RC1246

B

A

Repeater

☐ Coverage of repeater ☐ Coverage of station A ■ Coverage of station B

the relatively narrow spacing in frequency between transmit and receive. You need a very sensitive receiver working (often from the same antenna) with a transmitter only 600kHz away. To overcome this problem, considerable filtering is required, and repeaters will almost invariably employ cavity notch filters to reduce this de-sensitisation so the repeater does not gain the unenviable reputation of being 'deaf'.

The siting of a repeater is important, and often the major constraint for repeater builders after the filtering. Nowadays, prime radio sites are in great demand, and site owners can charge commercial rates for space. Fortunately, amateur repeaters were around long before cellular phones, and started before the explosion in mobile radio services in the last 10 to 20 years. This has meant that many amateurs have managed to secure favourable status with a number of mast owners. Besides, many licensed amateurs and repeater builders are themselves working in the industry.

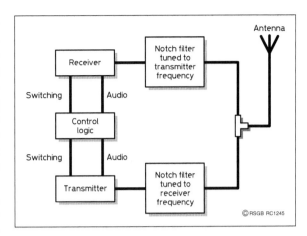

Fig 3. Block diagram of the 70cm repeater GB3BD in Bedfordshire: the receiver is at the bottom, logic circuitry is in the middle and transmitter is at the top

Despite all that, getting permission to use a good radio site can prove difficult, and more than one repeater has gone under because of site problems.

There are repeaters on many bands, from proposals to use 29MHz right up to 10GHz. There are also several specialist repeater projects. There is a pilot single-sideband (PSSB) repeater near Buxton in Derbyshire, and there are several FM television repeaters on the 1296MHz band. There are also linked repeaters and repeaters linked to the Internet, and CTCSS tone access is now gathering favour (but that is another subject!)

There has, regrettably, been a down side to repeaters as well. Being sited, as they have to be, on prime hilltop locations, they have sometimes attracted unauthorised use, and some repeaters, predominantly in urban areas, have suffered abuse over the years.

USING REPEATERS

Using repeaters involves *duplex* operation: you transmit and receive on different frequencies. Most available transceivers have this facility built in. There is a user's protocol. Firstly, repeaters are primarily intended for mobile or portable users. It is definitely unacceptable to 'hog' them for long lengths of time. Always welcome newcomers to join in, encourage membership of the group, and remember your normal amateur protocols are as valid on repeaters as on simplex operation (use of callsign, courtesy etc).

Before attempting to transmit, ensure that:

- Your transmitter and receiver are on the correct frequencies (remember the repeater split).
- Your tone access (if fitted) is operating correctly.

- Your peak deviation is set correctly (some repeaters will not relay your signal if it is incorrect). Any adjustments you have to make should be done into a dummy load, *not* on-air.

Avoid using the repeater from your base station; it is really intended for the benefit of local mobile and portable stations. If you really do intend to try it from a fixed station, use the lowest power to get into the repeater (under 1W is usual in the majority of situations where you can hear the repeater well).

Always listen before transmitting. Unless you are calling another specific station, simply announce that you are "listening through", eg "GM8LBC listening through GB3CS". On the other hand, if you are responding to someone specific, try something like, "GM0ZZZ from GM8LBC".

Once contact is established:

- At the beginning and the end of each over, you need give only your own callsign, eg "from GM8LBC".
- Change frequency to a simplex frequency at the first opportunity, especially if you are operating from a fixed station.
- Keep your overs short and to the point, or they may time-out, and do not forget to wait for the 'K' or 'blip' if the repeater uses one.
- Do not monopolise the repeater when busy as others may be waiting to use it.
- If your signal is very noisy into the repeater, or if you are only opening the repeater squelch intermittently, finish the contact and try again later.

JOINING UP

Repeaters have initiated many newcomers to the amateur radio hobby. Repeater outputs can be monitored easily, either with widely available amateur equipment, or perhaps with a scanner. Repeater groups themselves have been able to join up prospective users, and through local meetings, or contact, newcomers have been shown how to proceed. Many amateurs use repeaters in addition to their other amateur activities, perhaps using their local repeaters whilst travelling to and from work, then doing something entirely different in the evenings.

Amateur 'purists' have, particularly in the early years of repeater growth, shunned their existence, as not being truly in the 'ham spirit'. 30 years on, this is very much a minority view. In my personal case, I started with the hobby just as repeaters were starting. I received my licence in the mid-'seventies, just as the networks were in the early stages. I had my first QSOs on FM, and have been a member of various repeater groups over the years, worked though dozens, helped to build and maintain several, and now help to administer them on behalf of the RSGB.

It is the strength of our hobby that has presented so many opportunities for devotees to find their niches. Repeater builders are amongst the most technically competent and experienced in the hobby; many are employed in the PMR industry or in professional communications. Repeaters are often co-sited with major broadcasters or PMR users, so have to be of a high technical standard.

MORE INFORMATION

There is not a lot of additional reading about repeaters, but lists of the current UK networks are available in the annual *RSGB Yearbook*. A letter to the Chairman, RMG, c/o RSGB HQ will be directed to an appropriate member of the Repeater Management Group for reply.

It is also recommended that you should find out about your local repeater group and join it, thereby opening up many other possibilities for learning more about amateur repeaters. Your local repeater group may publish information, perhaps in the form of newsletters or information sheets.

The RSGB *Amateur Radio Operating Manual* has a whole chapter devoted to mobile and portable operation, including operation through repeaters.

Packet radio principles

Following the end of the Second World War there was a glut of surplus teleprinters on the market. Many of these machines (and their successors) were made by the venerable Creed company in Croydon. It wasn't long before these noisy, smelly and utterly lovable machines were being used for RTTY (radio teletype) operation on the amateur bands. Using an old Creed machine was preceded by a fair amount of construction of terminal units and modification or repair of the machine. This was the beginning of the world of data communications for the amateur. Prices were low and the satisfaction of working this mode was immense. As the affordable home computer came on the market, the old Creed machines were gradually replaced. Little did we know that an even bigger change was soon to take place, as the packet network was born and slowly became used world-wide. RTTY is still going strong of course – the modern computer and its versatile sound card have seen to that!

THE ORIGINS OF PACKET

Packet radio stems from a method (or *protocol*) used to set up communications links between computer networks. The protocol is known as *X25*. The amateur version had some changes made to it and is known as *AX25*, the 'A' standing for 'amateur'.

Packet radio is a method of data communication; it allows one computer to talk to another via radio. Obviously, there must be a way of connecting the digital signals in the computer (strings of 1s and 0s) to the transceiver. This is accomplished by a *terminal node controller* (TNC). This handles all communications between the computer and the transmitter and (in the reverse direction), between the receiver and the computer. Fig 1 illustrates this. In effect, the TNC is a type of modem. It takes the digital signals from the computer serial port (RS-232C) and turns them into sound signals that can be transmitted by the rig (Fig 2).

The TNC also handles the way in which the link works, sending control signals and instructions to the TNC at the other end. The only other thing to remember is that you will need some software to allow you to communicate with the TNC. There are many packages available, many of them shareware and freeware, so do look around until you find something you like. You may like to know that any 'terminal' program will allow you to send simple alphanumeric data to the TNC, but the specialist programs have many useful features included.

Fig 1. The constituent parts of a typical simple packet station

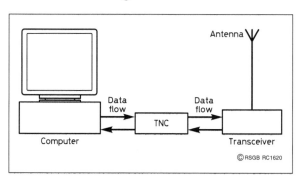

HOW IT WORKS

Before starting data transfer, the TNC must 'connect' with another station. This is done by issuing the connect command to the TNC, followed by the callsign of the station at the other end of the link. For example, to connect to G9ZZZ one would type 'Connect G9ZZZ'. This is normally abbreviated to 'C G9ZZZ'. Once the connection is made (the message 'CONNECTED TO G9ZZZ' appears on the screen), data transfer can begin.

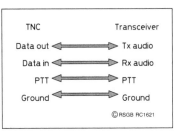

So how does packet work? The user types his message at the computer keyboard and the TNC assembles the data coming from the computer into little 'packets'. Each packet contains the callsign of the destination station, the callsign of the originating station, plus control and error detection information. This is in addition to the data actually being sent by the operator. The packet is then sent to the transceiver for transmission. If the data is received correctly, it is acknowledged and the next packet is sent. If the sent packet is corrupted, the receiving station will request a repeat. Theoretically the repeat requests could go on indefinitely but the number of repeats is set by the user, and normally 15 retries is the limit.

Fig 2. The TNC-to-transceiver link

There are three data transmission rates in use at the moment. These are 300bps (bits per second or baud), which is normally used on HF; 1200 and 9600 baud, used normally on VHF. This last speed normally requires modification of the transceiver without which the audio filters would not allow the signal to pass. There are some transceivers on the market designed to work at this speed without modification.

PACKET QSOs - *Contact with a station (Station On)*

How do you start a packet QSO? Well, in much the same way as you start a phone contact. Either call CQ, or arrange a QSO with a friend. Once connected to the remote station (remember the 'C G9ZZZ' command?) you can type text in at your computer, just as if you were talking to him or her. The responses from the other station will be displayed on the monitor for you to read.

A point about operating practice is worth mentioning here. When you have finished sending your latest sentence(s) use chevron signs (>>) to denote the end of a transmission. It is also important to remember to wait for the chevrons from the other end before you start sending, otherwise things can get a little confusing.

Most terminal programs send the text to the TNC at the end of a line. You can send it at any time you like by just hitting CR (carriage return).

A point about station identification when using packet: if your transmissions last longer than 15 minutes, you *must* send either a CW or phone ID. Please look in the *Amateur Radio Licence Terms, Provisions and Limitations Booklet BR68* for more information.

BULLETIN BOARDS

As well as connecting to another station and having a chat, it is also possible to connect to a *bulletin board* (its abbreviation 'BBS' stands for 'bulletin

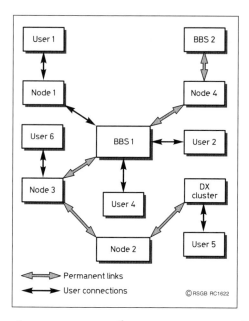

Fig 3. A typical packet network. A DX Cluster is a special sort of packet station set up to allow users to share up-to-the-minute DX information

board station' or 'bulletin board system'). A bulletin board can be connected to in the same way as a 'normal' station but most of the time there is no operator present. The sysop (system operator) provides a station that is connected to the packet network. The BBS stores messages from other amateurs. These messages can be either personal mail to a given station or they can be general messages (*bulletins*) intended for a general audience.

You will find all sorts of useful files on BBS computers. The files can be accessed and downloaded by your packet station. Mail for both you and for other stations can be sent to or from a BBS, and bulletins can be read (and you can get into some interesting debates at times).

There are two ways of checking for personal messages and seeing what bulletins have been posted. The first involves actually 'logging on' to the BBS. All BBSs will tell you when you log on if there are messages waiting for you. You can then read the mail when you want to. One thing here: please remember to 'kill' your mail once you finish with it. If you don't it will stay on the system, clogging up memory for no reason. You can imagine what would happen if we all just left our messages in memory. You can look at bulletins by scanning the BBS message list and downloading the messages you want directly. Alternatively, you can download the message list directly to your computer. Then you can decide which (if any) you want to read and then collect them later.

THE PACKET NETWORK

In one way, packet radio is probably one of the politest modes in the amateur service: when several stations are using the same frequency – as often happens – the various TNCs take it in turn to transmit. This reduces the amount of collisions (two stations transmitting at the same time). This will reduce the data transfer rate, as the number of transmitted packets is reduced, but this leads to greater spectrum efficiency.

One question you may be asking is "Where do all the messages come from?" Well, they actually come from other amateurs all around the world. It is not unusual to see messages from the USA or even further away. They travel here over *the packet network* (Fig 3).

The packet network is the 'Internet' of the amateur radio world. Each BBS is connected to the network, and hence to every other BBS. This allows them to access the fast links that run up and down the country so that messages can be passed throughout the UK. Some of these links are at microwave frequencies, others at UHF. Other countries have their own networks which work on the same principles as the UK one. All of these networks are linked together by HF and satellite 'gateways', allowing messages originating in the UK to travel to other parts of the world.

Packet radio may be found on most bands, but probably the most popular is 2m. Following the recent changes to the 144–146MHz band plan, digital modes have moved to the 144.800–144.990MHz sub-band. On HF, you will find digital modes just above the CW portions of the bands.

A word of warning: if you are using 300-baud packet on the HF bands, please avoid the beacon sub-bands, as interference can make beacon observation difficult.

YOUR FIRST PACKET STATION

Setting up a packet station need not cost the earth. I have seen TNCs that cost less than £70, free software, and an old IBM-XT or similar PC for under £20. You must remember that the PC is only being used as a terminal in this case. Later, you may want to move up to fax, AMTOR or SSTV, which will cost only a little more.

So how does it all go together? You will need to make (or buy) a cable to go between the PC and the TNC. Make sure that you get the right number of pins. Some computers have RS-232C outputs with nine pins, while others have 25 pins (you will have to look). Once you've linked the PC to the TNC, you must make up a lead between the microphone input and audio output sockets of the rig to the audio output and audio input sockets, respectively, on the TNC. Once this is done you are ready to use packet. Just load the software, fire everything up and you're ready to go – or almost.

Some TNCs have parameters that must be set up. Here, I am afraid that you will need to read the instruction manual to see what you need to do. You may well find that the manual gives you a lot of information, so it is well worth reading through. However, all TNCs have a 'default' set of parameters which should work first time. You can refine their values once you become knowledgeable about their effects.

Before you rush off, buy the kit and have a go, can I make a suggestion? There are several good books on the market about packet. Look at the books listed below and you will find all the help you need. The British Amateur Teledata Group (BARTG) is also a good information source, and a contact address is given.

Good luck, and enjoy your packet operations.

REFERENCES
[1] *Your First Packet Station*, G0WSJ, RSGB.
[2] *Packet Radio Primer*, G8UYZ and G8NZU, RSGB.
[3] BARTG membership enquiries: Bill McGill, G0DXB, 14 Farquhar Road, Maltby, Rotherham S66 7PD. E-mail: members@bartg.demon.co.uk.

Using 10GHz

Mention microwaves and many beginners will say "It's too complicated, too costly and too difficult for me". Well, present-day, state-of-the-art, narrow-band equipment designs almost certainly are too difficult and possibly too costly for beginners. Many newcomers do, indeed, appear to be put off by this high technology.

So where do beginners start? This article may help to revive interest in the simple, easy, and inexpensive wideband FM (WBFM) designs that most of today's advanced operators used before the latest state-of-the-art designs became available for experienced home constructors.

The ideas are not new, so don't expect a 'blow-by-blow' description. Full details are available in the references quoted. The intention is to encourage you to get hold of this information, have a go, produce working equipment, get the feel for microwave operating and then, perhaps, to have a go at some of the more advanced microwave designs as your skills and knowledge improve with practice. Believe me, you *don't* have to be a skilled constructor or operator to take the first steps.

WHAT'S NEEDED?

You will need the following major 'ingredients' to construct a 10GHz WBFM transceiver:

- An antenna (see, for example, the horn antenna in reference [2], pp34–41)
- A mast (for fixed station working) or tripod (for portable working see reference [2], pp107–108)
- An in-line Gunn oscillator/mixer Doppler intruder alarm module (see reference [1], pp98–102)
- A Gunn oscillator power supply/modulator module (see reference [1], pp94–97)
- A low-noise, wide-band IF preamplifier module (see reference [2], pp122–124 or pp125–127)
- An FM receiver to tune the chosen IF (see, for example, reference [2], pp126–136 – there are many alternatives, see later)
- Various screened (metal) boxes, switches, plugs and sockets, controls and control knobs, cables, wire and general hardware – see later
- A 12V power supply at 1.5A maximum (including a scanner receiver), battery for portable use, mains for fixed station use.

So, how does it all work?

THE RECEIVER

A basic superheterodyne receiver for *any* frequency consists of two main parts, as shown in Fig 1. The first is a *front-end converter* consisting of a local oscillator (LO) and a mixer to produce a lower intermediate frequency (IF), the sum or difference of the LO frequency and the received frequency. The reason for converting the high-frequency received signal down to a lower frequency is because it is easier to amplify, filter and demodulate.

The second part is a *back-end*, consisting of IF amplifiers, filters, often a second oscillator/mixer (to convert signals to a second, even lower IF), a demodulator and an output stage.

A 10GHz receiver is no different. It would be very difficult for a beginner to make a microwave oscillator and mixer. Fortunately, there are ready-made oscillator/mixer units available in the form of surplus 'in-line' Doppler intruder alarm units. These are obtainable from most amateur radio rallies for £5 or £10. They consist of a Gunn device oscillator and a diode mixer mounted in-line inside a short length of (usually) waveguide 16 (WG16) which forms *cavities* (tuned circuits). WG16 is the standard waveguide size for 10GHz (3cm).

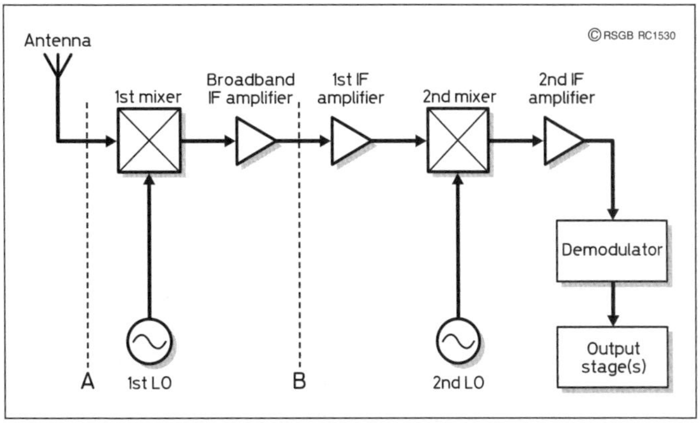

Fig 1. Block diagram of a basic double-conversion super-heterodyne receiver. The front-end converter is between A and B, and the back-end is all to the right of B

An antenna is connected to the open end of the waveguide and the received signal collected by the antenna travels down the waveguide to the mixer diode. LO input to the mixer is via a hole or slot in the end wall of the LO cavity. The IF output is taken from the mixer diode terminal on the Doppler module and connected to a suitable receiver back-end.

The back-end of the receiver can be almost anything you like, provided that it will tune to an IF somewhere between 10.7MHz (minimum) and, say, 50 or 60MHz, and has the necessary amplifiers, filters, demodulator and output stages. IFs below 10.7MHz can't be used because of LO noise, while IF signals above about 60MHz may be bypassed by the mixer diode decoupling capacitor built into the Doppler module. At one time a favourite choice was an FM broadcast band receiver covering 88 to 108MHz: these days there are so many FM stations in that band that it is almost impossible to avoid IF breakthrough.

Several choices can be made: a complete 10.7MHz back-end (see reference [2], pp126–136), a simple 50MHz receiver (see reference [2], pp74–86), or a scanner tuned to whatever IF seems to work best. The choice is yours.

THE TRANSMITTER

The output of the Doppler module Gunn oscillator is usually in the range 10 to 20mW. The mixer diode uses only a small proportion of this, leaving the remainder to escape from the open end of the Doppler module which of course is connected to the antenna. Since both the receiver input and the transmitter output (the open end of the Doppler module) are connected to the antenna, we don't need to switch the antenna between transmit and receive.

All we need in order to use the Gunn oscillator as a transmitter is some means of modulating it with speech or audio tone.

By now, you can probably see that the heart of a simple 10GHz transceiver is the Gunn oscillator used both for receive and transmit. The Gunn oscillator needs a stabilised, variable, low-voltage bias power supply (typically 6 to 9V at about 150mA) to operate. If you want to know exactly how a Gunn device works, refer to reference [3]. Two factors determine the exact Gunn oscillator frequency – the size of the cavity (tuned circuit) in which it is mounted and the bias voltage which is applied.

The cavity can be tuned through several hundred megahertz by means of a metal screw in the cavity wall near to the Gunn device. This is a useful way of coarse tuning; inserting the screw into the cavity lowers the oscillator frequency, while withdrawing it raises the frequency.

Varying the Gunn bias voltage also changes the oscillator frequency. This is an effect known as *frequency pushing*. Lowering the voltage decreases the frequency, while raising it increases the frequency. The effective tuning range is limited to a few tens of megahertz, so that this is a very convenient way of 'fine tuning' the oscillator. A few millivolts of audio or tone modulation can be added to the bias voltage to produce high-quality FM modulation.

The Gunn bias (power) supply therefore needs to be variable but, once set to a particular voltage, must be stable and free from noise. An unstable voltage or a noisy supply will result in an unstable and noisy oscillator! A very reliable Gunn supply with modulation facilities was described in reference [1], pp94–97. Use a 10-turn potentiometer and turns-counting dial for the 'fine-tune' control – this gives good bandspread and frequency-setting accuracy and is well worth the extra cost. The turns-counting dial can be calibrated in frequency to make frequency setting easier.

OTHER CONSIDERATIONS

Two other things need to be considered. Problem number one: coaxial feeder can't be used between the antenna and the Doppler module because the feeder losses at 10GHz would be astronomical! Problem number two: the IF output level of the mixer is very low so, again, losses in any connecting cable are not acceptable. What's to be done?

Both problems can be quite easily resolved by building the transceiver in two units. Mount the Doppler module and wide-band amplifier in a waterproof, screened (metal) box, attach the antenna directly to the same box and mount the whole of this 'microwave head' at the top of a mast for fixed station use – see reference [1], p94, or reference [2], p126 – or on a tripod for portable use.

Build the Gunn power supply/modulator and all the controls (including the receiver back-end) into a base unit which can be used in the shack for fixed-station use or in the car for portable use.

Of course, if you want to work both fixed and portable, you will want to leave the mast-mounted microwave head at home and use a second, identical head for portable use. Build two masthead units or build another Doppler module into the base unit and use simple switching to change from the 'local' to the 'remote' microwave module, as suggested in Fig 2. S1 is the power on/off switch, S2 selects speech/ tone, J1 is the mic jack, J2 the key jack, J3 coax to external receiver, J4 coax to mast-head unit, SK1 is for 12V in, SK2 for 12V to external receiver, P1 is the Gunn 'fine tune' control – 10 turn potentiometer/ turns counter. The additional control S3 switches internal/external Gunn.

Built like this, one ordinary coaxial cable can now be used to connect the base unit to the remote unit, sending the modulated Gunn bias up to the remote microwave head and the amplified (broad-band) IF down from the remote microwave head to the base unit.

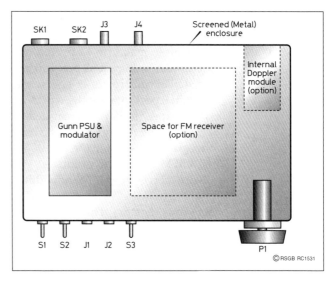

Fig 2. A suggested layout for a base unit, extended to include switchable internal Gunn

Full duplex operation was mentioned (see reference [2], p125). Unless the two stations share an identical IF, this will not be possible and it is probably the exception rather than the rule. A simple addition, shown in Fig 3, allows the G4KNZ Gunn PSU/modulator to be set for fixed-frequency transmit, while allowing receiver tuning over the whole range of frequency pushing.

When everything appears to be working correctly, the last things to be done are to retune the Doppler Gunn oscillator from its usual ISM (Industrial, Scientific and Medical) frequency of 10,687MHz into the amateur band, somewhere between 10,370 and 10,400MHz, which is the current UK WBFM sub-band. To do this you'll need the help of someone with a 10GHz counter or high-Q wavemeter. If all else fails, try one of the Microwave Round Tables; dates and venues are announced in the 'Microwave' column in *RadCom* and in the *Microwave Newsletter* [4].

Setting up a Gunn oscillator is described in reference [2], pp134–135. Setting up the G4KNZ module, as modified in Fig 2, is very similar. You should have an oscillator tuning range of about ±10MHz. The receiver will respond to both images, that is, $f_{osc} \pm F_{IF}$. If the IF is 10MHz, this means that the receiver will tune from 10,370MHz minus 10MHz to 10,390MHz plus 10MHz, ie from 10,360 to 10,400MHz. Using either image means that you can tune both the narrowband segment (10,368 to 10,370MHz) and most of the wide-band segment (10,370 to 10,410MHz). You might be lucky enough to hear a beacon which will provide a very accurate frequency marker to check your calibration.

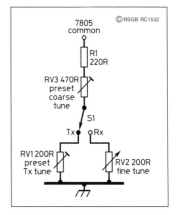

Fig 3. Modifications to G4KNZ Gunn PSU/modulator to give fixed transmit frequency and tunable receiver

CONCLUSION

It is possible to construct an entirely practical 10GHz WBFM transceiver using ready-made 10GHz Doppler oscillator/mixer modules, with some kind of FM receiver back-end. The overall cost of the whole project will depend on how much you like to experiment, how elaborate a transceiver you want to make, how carefully you shop around for parts and, to some extent, on how big a junk box you have! The point is that you do not have to be a skilled constructor or operator to get results.

Don't expect to work the world with simple, low-power 10GHz WBFM, but do expect to have a great deal of fun and gain useful experience and skills in trying to work your pals, either from home or portable (over much longer distances) using the 10mW or so transmitter power that typical, simple WBFM equipment produces.

Since there's plenty of bandwidth available, Doppler modules can be used for amateur fast-scan television (ATV) or very high-speed packet links – but that's another story. Think about it – microwaves have a lot to offer.

REFERENCES
[1] *Practical Transmitters for Novices*, GW4HWR, RSGB.
[2] *Practical Receivers for Beginners*, GW4HWR, RSGB.
[3] *Microwave Handbook*, Vols 1, 2 and 3, ed G3PFR, RSGB.
[4] *Microwave Newsletter*, eds G3PHO and G8AGN, issued 10 times per year.

Decibels

When faced with calculations involving decibel (dB) functions, many people will suddenly be possessed of a desire to mow the lawn, take up flower arranging or iron a dozen or so shirts! In fact, decibel notation is the ideal way to understand the relationship between two different values – regardless of what these values are. The purpose of this article is to explain how decibels are calculated and used (and sometimes misused!).

WHAT ARE DECIBELS?

The first – and probably the most important – thing to establish is that 'dB' on its own means nothing. Try clapping your hands with one tied behind your back; unless you are an octopus it cannot be done. You need both hands to carry out the task, since you require one hand to strike the other. Equally, decibel calculations require the decibel to *relate* to something, otherwise any calculation is meaningless. This is why RF output is often shown as dBW – the decibel has, in this case, been established as *relative to 1W*. However, decibels could, in theory, refer to absolutely anything.

For example, let's start with decibel 'tea-bags' as a reference. Since an increase of 3dB equates to doubling the value (I'll explain why later), increasing 'tea-bags' by 3dB doubles the number of tea-bags. I agree that this example is strange, unless you are into homebrew equipment [Aarrgh!!! – *Ed*] but the same principle applies to power output. Increasing 10W by 3dB gives you 20W. Equally, decreasing 10W by 3dB will halve your power to 5W. In the real world, once the subject matter has been established, the reference value (whether 'watts', 'tea bags', or anything else) is often dropped. So the specification for a linear amplifier may refer to a 20dB gain; this would mean that the amplifier's output is 100 times its input – since we are obviously looking at power, the correct form of 'dBW' is shortened to 'dB'.

To summarise – for a quantity expressed in decibels to possess a real value, it must relate to a known quantity (eg dBW).

The reason for using decibel notation is that it establishes a relationship between two values – in other words, it is a *ratio*. This is important, because fixed-value calculations can potentially misrepresent what is actually happening. For example, increasing the power output of a transceiver from 2W to 4W has doubled its output (+3dB). However, increasing the output of a 100W transceiver by the same amount (ie 2W) only increases the power by a paltry 2% (+0.086dB). To achieve the same 'relative' improvement, the transceiver's output would need to increase to 200W.

SHORTCUTS

Fortunately there are some shortcuts which can make life much easier. Since dBW is, by definition, dB relative to 1W, a value of 0dBW is equal to 1W (ie zero change).

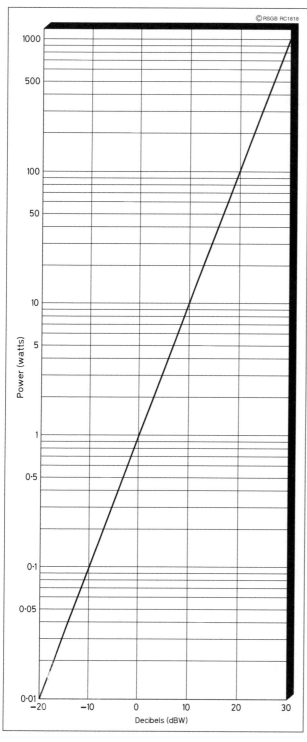

© RSGB RC1818

Fig 1. At-a-glance graph for converting between power and decibels

The first shortcut is – *multiply* the power by 10 for every 10dB *added*. This is an easy one to remember. From this you will see that *adding* 20dB will *multiply* your power by 100; if you start with 4W and increase by 20dB, you end up with 400W. Going the other way, *divide* the power by 10 for every 10dB *subtracted*. For example, 10dB *below* (ie less than) 400W is 40W.

The second shortcut is – *double* the power for every 3dB *added*, and *halve* the power for every 3dB *subtracted*. For example 60W is 3dB *above* 30W, and 15W is 3dB *below* 30W. This is not 100% accurate, but close enough for all but the purists.

The third shortcut is – *multiply* the power by 1¼ for every 1dB *added* and *divide* the power by 1¼ for every 3dB *subtracted*. Once again, this is not 100% accurate, but close enough.

APPLICATION

The maximum legal power limit on most bands is +26dBW. If 0dBW is 1W, then +20dBW is 10W. Double this to 200W for +23dBW, and again for +26dBW, giving the 400W maximum which we all know, love, and wouldn't dream of exceeding. Fig 1 provides a graphical method of conversion and Table 1 provides a summary of the shortcuts.

CAVEAT EMPTOR!

Unless you have at least a general understanding of decibel calculations, it is very easy to be misled by some manufacturers' claims about gain in decibels as related to antennas.

Thankfully, deliberate misrepresentation is rare within the amateur world but the occasional undefined decibel gain figure is not unknown. The 'gain' of different types of antenna will often determine their sales potential, since we would all like minute antennas with huge gain characteristics.

The gain figures for antennas are usually specified in dBi or dBd. dBi relates to gain relative to a theoretical 'isotropic' antenna, which is one which radiates equally in all directions; the second relates to gain relative to a 'dipole' which doesn't radiate equally in all directions. A dipole's maximum gain is in a direction perpendicular to its elements, and is some 2.1dB better than an isotropic antenna, so we can say that a 2m 9-element Yagi having a gain of 9dBd also has a gain of 11.1dBi. A specification given in dBi can, therefore, give the impression of greater gain.

When purchasing antennas, if the manufacturer indicates gain figures in dB only, ask if they are dBd or dBi – you may save yourself an expensive disappointment.

Table 1. Shortcuts for converting between power and decibels

dB	Mathematical relationship
+30	One thousand times
+20	One hundred times
+10	Ten times
+3	Twice
+1	One and a quarter times
0	No change
−1	Four fifths (= 1 ÷ 1¼)
−3	One half
−10	One tenth
−20	One hundredth
−30	One thousandth

Antenna maintenance

Spring is the ideal time to do something about those antennas that have suffered during the extremes of the winter months. This work involves maintenance of existing antennas and the installation of better ones.

ANTENNA UPKEEP

We all know that a length of wire will radiate but a length of wire will radiate even better if it is in good condition.

Most antenna losses are caused by corrosion, which increases the resistive losses in the antenna. When maintenance is carried out on an antenna system, all mechanical joints should be dismantled and corrosion removed. They can then be coated with grease before reassembly. Soldered joints should be inspected and remade if they look suspect. Insulators, particularly at high voltage points (such as at the ends of dipoles or long wire antennas), should be cleaned.

Check that the transmission lines (coaxial cables or twin feeders) are in good condition and that the connectors are free from corrosion. They, too, should be coated in grease after cleaning and before re-assembly.

IMPROVING THE ANTENNA

In general, the most practical way of improving the performance of an existing antenna is to erect it as high as is practicable at your location. An antenna that is low and close to the house is also close to the electrical QRM which envelopes our dwellings as shown in Fig 1(a). If the antenna can be raised it will be further from this interference, which will enable us to hear stations that would otherwise be hidden in the electrical noise; see Fig 1(b).

The antenna will also be further away from TVs and other domestic electronic equipment and the chances of TVI are reduced. Also, when the height of the antenna is increased, the radiation pattern favours DX stations to a greater degree (because of the lower angle of radiation).

A worthwhile improvement can often be made with a modest increase in

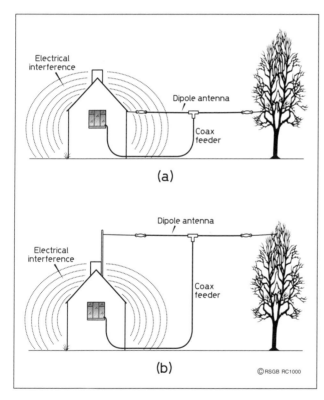

Fig 1. (a) Dipole antenna at a low height, showing it in the strong electrical interference field. (b) Antenna height is increased, thus reducing the electrical interference and increasing the signal

height, such as moving one end of the antenna from the eves of the house to the chimney or connecting it to a higher branch of a tree.

TRANSMISSION LINES (FEEDERS)

Although the antenna is installed as high as possible, the transceiver is installed at a convenient place indoors, where it is readily accessible and out of the weather.

For the best signal-to-noise ratio, and for avoiding TVI, it is better if the connector between the antenna and the transceiver does not radiate or receive signals. This

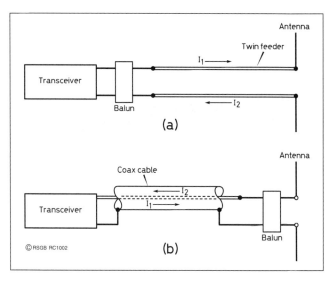

is achieved using a transmission line, which can comprise coaxial cable ('coax') or twin line feeder.

The twin-feeder form of transmission line comprises two conductors placed in parallel and close together as shown in Fig 2(a). The current flowing in the two conductors travels in opposite directions; in other words they are 180° out of phase with each other. If the two currents also have equal amplitudes, the electromagnetic field generated by each conductor will cancel that generated by the other, and the line will not radiate or receive radio energy.

With coax (Fig 2(b)), the current flowing in the outer conductor does so on the inner surface so that radiated or received RF energy is cancelled, just the same as in the twin-line feeder. In addition, the outer conductor of coax acts as a shield, confining the RF energy within the line. Because of its construction, coaxial cable is said to be *unbalanced*.

If the currents flowing in the two lines of the feeder are not equal in amplitude or not exactly 180° out of phase, the line will radiate or receive RF energy.

Centre-fed dipoles and loops are *balanced*, meaning that they are electrically symmetrical with respect to the feed point (where the feeder is connected to the antenna). A balanced antenna should be fed with a balanced feeder system to preserve this electrical symmetry with respect to ground. However, the antenna connector on the back of the transceiver is for coaxial cable, and is thus unbalanced. Some method of method of connecting the transmission line to the antenna without upsetting the symmetry of the antenna itself is required. A device for converting a balanced circuit to an unbalanced circuit is a *balun* (a contraction of 'balanced to unbalanced'); see Fig 3.

Coaxial cable has the advantage of being very practical for

Fig 2. Showing the field-cancelling effects of current in (a) twin feeder and (b) coaxial cable

Fig 3. (a) Circuit diagram of a 1:1 balun. (b) Its physical construction

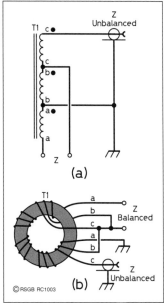

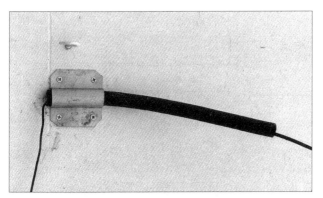

75Ω twin-feed supported in a length of plastic tube and fixed to the wall with a metal bracket

most amateur radio installations. Because of the excellent shielding afforded by its outer screening, coax can be run up a metal tower or taped together with numerous other cables with virtually no interaction. At the top of a tower, coax can be used with a rotating beam without shorting or twisting conductor problems. Coax can even be buried underground in plastic tubing.

The disadvantage of it is that it normally requires some matching unit or balun at the antenna. Also, precautions have to be taken to ensure that the connection is absolutely watertight. If water gets into the tube-like structure of coax then the cable is ruined.

Coaxial cable is relatively heavy compared with a single copper wire. This means that there is a fair amount of mechanical stress on the coax-to-antenna connection of a centre-fed dipole, especially when using heavy-duty coax.

Open-wire feeder must be kept away from metal objects by several times the spacing between its conductors. Despite this mechanical difficulty, there are often compelling reasons for using this type of feeder. One of these is the low loss incurred when using twin-line feeder in a multi-band antenna. In addition, twin feeder is far cheaper than coaxial cable and it is much lighter.

The 72Ω twin feeding my quad is now six years old and has not yet failed. I used to change my coax each spring!

An excellent and cheap insulator for HF antennas is a 1m long length of heavy monofilament fishing line or strimmer cord, but learn to tie a bowline or fisherman's knot otherwise it will easily come undone!

Baluns

The prime purpose of a balun (a contraction of *bal*anced-to-*un*balanced') is to allow an unbalanced source to drive a balanced load or vice versa. Some types of balun will also yield an impedance transformation but this should be regarded as a secondary function.

BALANCED SYSTEM

Before getting into the details of baluns, it is necessary to understand just what is meant by a balanced load, and why feeding such a load from an unbalanced source can create problems.

Fig 1 shows a typical balanced load. The arrangement is symmetrical about the centre line. Each point on the left-hand side is mirrored by an equivalent point on the right-hand side, where the currents and voltages are equal in amplitude but opposite in phase.

In the dipole itself, the currents in the two legs create fields which add together to generate the usual 'figure of eight' radiation pattern. The fields generated by each half of the feeder, though, cancel out each other so that there is no radiation from it.

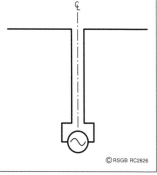

© RSGB RC2626

Fig 1. The standard dipole is electrically symmetrical about the centre line

FEEDING VIA COAX

Now consider the same dipole fed through a length of coaxial cable from a typical transmitter, as shown in Fig 2(a).

The current flowing in the inner conductor of the cable has only one destination, the left-hand leg of the dipole. That flowing in the outer of the cable, however, has two destinations – the right-hand leg of the dipole and back down the outside of the cable to ground.

Fig 2(b) shows a somewhat simplified representation of the various current paths, with I_3 being that flowing back down the outside of the coaxial cable. The result of having a path for I_3 is that a top-fed vertical antenna is in effect put in parallel with the right-hand leg of the dipole. This vertical antenna will of course radiate.

The amplitude of I_3 is dependent upon the length of cable being used. If it is an odd multiple of a quarter-wave, the feed

Fig 2. If coax is used to feed a dipole, this generates a current path to earth, effectively resulting in a top-fed vertical antenna

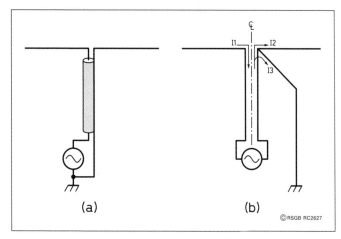

(a)   (b)

© RSGB RC2627

impedance of the effective vertical antenna is very high so I_3 is low. Under these conditions the unbalancing effect on the dipole is insignificant.

If, though, the cable is a multiple of a half-wave, the feed impedance is low and I_3 is therefore high. As would be expected, the unbalancing effect is also high.

POTENTIAL PROBLEMS

Having explained the situation, the obvious question to be answered is: does it really matter? In order to answer this question, we need to consider two separate aspects: first, the effect on the antenna's ability to radiate; second, what might be called the *system effects* – such matters as EMC, VSWR measurement and so on.

Feeding a normal wire dipole via coax causes some change to the radiation pattern but this may not be particularly significant. There can in fact be some benefit, in that radiation from the vertical antenna represented by the outer of the coax fills in to some extent the nulls off the ends of the dipole. Should the dipole be part of an array, a Yagi beam for instance, the effects of feeding directly via coax can be quite significant. The forward gain will be reduced, as will the front-to-back ratio, and some side lobes will also appear. Similar problems will also occur with other forms of array based on balanced radiators such as the cubical quad.

Of the system effects, EMC is probably the one of greatest concern. In normal installations the coax will enter the shack and be in close proximity to house wiring. Radio-frequency energy radiated from the outer of the coax is highly likely to find its way into all the surrounding wiring and cause breakthrough problems.

Under normal circumstances, the VSWR on a transmission line is dependent only upon the load impedance and the characteristic impedance of the line. Changing the line length should not cause the VSWR, as indicated by the usual VSWR meter, to change. Sometimes, though, it is found that the VSWR reading *does* change quite significantly with line length. This variation of reading is indicating that there is something odd about the set-up. Feeding a balanced load via coax represents one of the possible oddities.

Fig 2(b) shows that the actual load is made up of the dipole plus a top-fed vertical antenna. Also, as mentioned previously, the level of I_3 is dependent upon the line length. In effect, then, the load impedance at the antenna terminals varies with line length. With VSWR being dependent upon this load impedance, varying the line length will cause the VSWR meter reading to change.

It is also possible that I_3 flowing back down the outer of the coax will affect the operation of the VSWR meter, thus causing an additional source of confusion.

So, having established that feeding a balanced load via coax causes the radiation pattern of the antenna to change, the EMC threat to increase and the measurement of VSWR to become somewhat unpredictable, the next question is: can anything be done to improve the situation? Fortunately the answer is 'yes', and one of the solutions is to use a balun.

NARROW-BAND BALUNS

There are many types of balun, each having its advantages and disadvantages. Let's look first at some narrow-band arrangements – these being suitable for single-band antennas.

Fig 3(a) shows a very simple arrangement using an additional half-wavelength of coax. One feature of a half-wave line is that the voltage at the output is equal in amplitude to that at the input but with the phase reversed. With the arrangement shown, the voltages at A and B are equal in amplitude but of opposite phase. The voltage between A and B is twice that of the input. As a result, the load impedance for a 50Ω input impedance must be 200Ω.

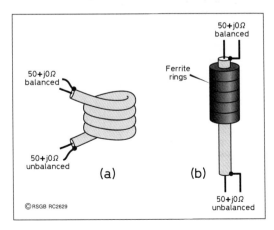

It should be noted that the antenna is not connected to the outer of the coax. If the inner of the coax does not have a DC path to earth, there can be problems with static build-up, particularly when there are electrical storms in the locality.

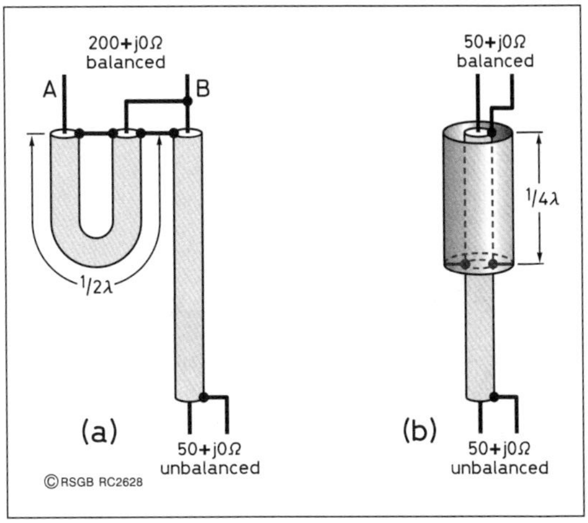

If a 1:1 impedance transformation is required, the arrangement shown in Fig 3(b) will suit the need. This uses a quarter-wave sleeve effectively to decouple the last section of the coax. Note that the top end of the sleeve needs to be well insulated and the bottom end connected to the outer of the coax.

Although these two baluns have the virtue of simplicity, they are only really suitable for the higher frequencies owing to the lengths of line needed to make them.

Fig 3. Narrow-band baluns can be made quite simply. A half-wave of coax, connected as in (a), generates a 180° phase shift. A quarter-wave sleeve, (b), effectively decouples the last part of the coax

BROAD-BAND BALUNS

These can be one of two basic types: those which force the currents in the two halves of the antenna to be equal in amplitude but of opposite phase and those which force the voltages to have this relationship. If the antenna is truly balanced, both achieve the same effect. One problem with wire antennas at the lower frequencies is that it is difficult for many reasons to achieve a fully balanced arrangement. The current balun ensures that in such cases the currents in both conductors of the feeder are equal in amplitude.

Fig 4. Two simple current-mode baluns. Both use inductance to reduce the current on the coax outer

Fig 4 shows two simple types of current-mode balun, both of which provide a 1:1 impedance ratio. These work on the principle of providing an impedance to restrict the flow of an out-of-balance current. In the case of Fig 4(a) this is achieved by coiling up the coax near to the feed point of the antenna. The coil has no effect on the normal signal flowing up the coax but looks like an inductance to any current trying to return via the outer. The same effect

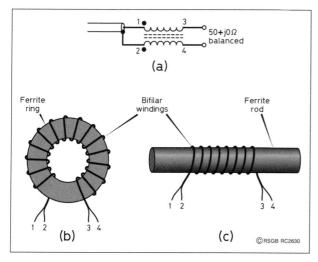

Fig 5. A bifilar winding on a ferrite core, either a toroid or a rod, yields a simple current-mode balun. Note that the dots indicate the starting points of each winding

is achieved in Fig 4(b) by threading ferrite rings over the coax.

One difficulty with the arrangement shown in Fig 4(a) is that it can be difficult to make it work effectively over a wide frequency range – say, 3.5 to 30MHz. Providing sufficient turns to cope with 3.5MHz is likely to result in the interwinding capacitance being too high for effective operation at 30MHz. The situation can be improved by winding the turns on a ferrite rod or ring.

A current-mode balun can also be constructed by using a bifilar winding on a toroid or ferrite rod, as shown in Fig 5. It must be understood that these bifilar windings act like transmission lines, which can limit the performance of the arrangements in some circumstances and yield rather erratic results. Ruthroff [1] advocated the addition of a third winding to the simple bifilar winding of Fig 5(a) to yield the arrangement shown in Fig 6(a). Although this third winding overcomes some of the problems of the two-winding arrangement, it has the effect of turning the balun into a voltage-mode device. Windings 1-3 and W3 act like an auto-transformer, so that the voltage at point A is half that of the input. The voltage at point B will also be half that of the input but with a phase reversal. This arrangement tends to be regarded as the 'standard' for 1:1 impedance ratio baluns.

The simplest form of voltage-mode balun, albeit with a 4:1 impedance step-up, uses the same basic arrangement as Fig 5(a), but with the windings connected in a different way, as shown in Fig 6(b). Construction is as for the examples in Fig 5(b) and Fig 5(c).

There are two (often conflicting) criteria associated with the design of voltage-mode baluns. First, the inductive reactance of the windings should be high; second, the leakage reactance should be low compared with the load impedance in each case. The first usually determines the low-frequency limit of operation whilst the second determines the high-frequency limit.

Fig 6(a). Adding a third winding, W3, to the simple current-mode balun yields a 1:1 voltage-mode balun. (b) Connecting the two bifilar windings in a different way gives a 4:1 impedance step-up design

CONCLUSIONS

This has been a fairly brief introduction into the subject and has (quite deliberately) begged the question of which design is 'best'.

The reason for this omission is that the use of baluns tends to result in compromises having to be made – what works well in one application may be a total failure in others. The sensible approach is to try a few different ideas and select that which gives the best performance in your particular set-up. Fortunately,

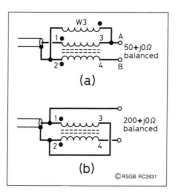

the components used are reasonably inexpensive and can easily be recycled for the different arrangements.

For those wanting to have a go, references [2–7] list articles and books containing more details of the different arrangements. You might be bemused by the fact that some authors will be enthusiastic about a particular arrangement whilst others regard it with horror. Take due note of any objections to the various designs but do not let these put you off trying them.

REFERENCES

[1] 'Some broad-band transformers', C L Ruthroff, *Proc IRE*, Vol 47, August 1959.
[2] 'Balance to unbalance transformers', Ian White, G3SEK, *Radio Communication* December 1989 (highly recommended reading).
[3] *Radio Communication Handbook*, RSGB.
[4] *ARRL Handbook*, ARRL.
[5] *HF Antennas for All Locations*, Les Moxon, G6XN, RSGB.
[6] *Backyard Antennas*, Peter Dodd, G3LDO, RSGB, 2000.
[7] *Transmission Line Transformers*, ARRL.
[8] *Reflections, Transmission Lines and Antennas*, ARRL.

How the cathode-ray tube works

The cathode-ray tube (CRT) is the most flexible measuring instrument in the electronics laboratory. Its construction is relatively simple, and has remained essentially constant since its invention by Braun in 1897. A selection of some different types and sizes is shown in the photograph.

DESCRIPTION

The basic oscilloscope CRT is shown in Fig 1, using its standard circuit symbol, which is very similar to its actual construction. Very much an electronic 'ship-in-a-bottle', all of its electrodes are contained inside an evacuated glass envelope.

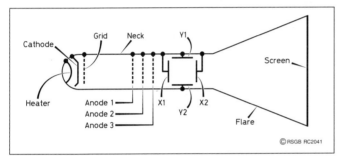

Fig 1. Circuit symbol for the cathode-ray tube

Electron emission

The *cathode* emits a copious supply of electrons when raised to red heat by the *heater*. Only a minute fraction of these are allowed past the *grid*, a metallic cylinder capped at one end with a plate containing a circular aperture. This electrode is maintained at a potential which is negative compared with that on the cathode, and tends to repel the electrons back towards the cathode. Those electrons which pass beyond the grid are thus controlled by the potential on the grid, which is varied by the 'brilliance' (or 'intensity') control on the front panel of the oscilloscope.

Anodes

Beyond the grid (at a distance much greater than that between the cathode and the grid) are three more circular plates with small apertures in their centres. These are called *Anodes 1, 2* and *3* (working away from the grid) and have very special roles to play.

Anode 1 and Anode 3 are usually connected together and are maintained at a potential of between +600V DC and +5000V DC relative to the cathode; the exact voltage depends upon the design of the equipment operating the CRT and its purpose. This extra high tension (EHT) attracts those electrons which have passed beyond the grid and accelerates them to very high speeds (comparable with the speed of light – see later). The beam of electrons diverges slightly as it approaches the anode structure, and is travelling so fast that it passes through the holes in the three electrodes, rather than colliding with them. Anode 2 is maintained at a lower positive potential than Anodes 1 and 3 (usually around +150V DC), and its purpose is to 'bend' the electron

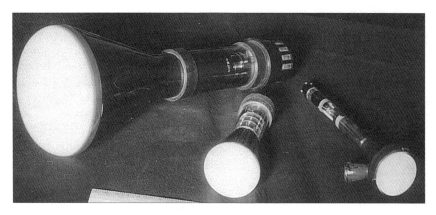

beam so that it converges rather than diverges. The amount of convergence is controlled by the potential on Anode 2, this being varied by the 'focus' control.

The electrodes considered so far form two groups within the cathode-ray tube. The heater/cathode/grid combination is known as the *electron gun*, as its purpose is to 'fire' electrons towards the anodes. The three anodes are known collectively as the *electron lens*, and are required to focus the electron beam on to the screen. Turning the focus control changes the focal length of the lens.

Deflection

Beyond the electron lens are two parallel pairs of *deflector plates*, as shown in Fig 2, which is *not* to scale. First come the *Y-plates*, two horizontal plates between which a potential can be established; this will exert a transverse force on the beam passing between them and thus will deflect the beam up or down (ie along the y-axis), depending upon the polarity of the voltage applied. The *X-plates* then perform an identical function except that the deflection produced is horizontal (along the x-axis).

Display

Once clear of both pairs of deflector plates, the beam continues on its way to the screen at a constant speed. The screen is simply a coating of phosphor powder on the inside of the glass envelope at the end of the *flare*, where the tube

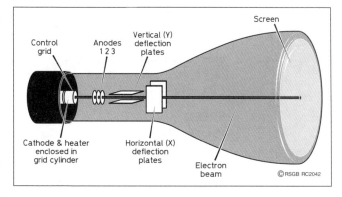

Fig 2. The electrode structure within the cathode-ray tube

increases in diameter between the neck and the screen, as Fig 2 shows. The purpose of the phosphor is to glow under the bombardment of electrons – the more intense the bombardment, the brighter the glow. Different phosphor materials give different colours, and the simplest of phosphor materials is zinc sulphide, which glows green.

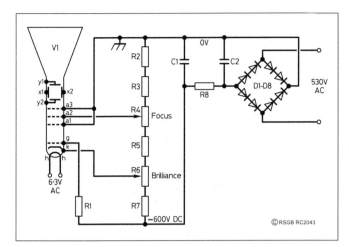

Fig 3. A simple circuit for biasing the electrodes of a CRT

Connections

The cathode-ray tube is a form of thermionic valve and, in common with its more numerous brethren, has the connections to its electrodes brought out to a series of pins in the tube base. This mates with a socket, to which the connections are made from the rest of the equipment of which the CRT is a part.

If it is assumed that no electrons strike the electron lens on their way through, there is no circuit present, as there is no conventional anode to collect the electrons. To overcome this problem, and to prevent the screen charging up, the insides of the neck and flare are usually coated with colloidal graphite which, in turn, is connected to Anode 3. This collects the secondary electrons emitted when electrons strike the screen, thus completing the circuit.

REFINEMENTS

There are two main changes to the structure which are found in more sophisticated CRTs.

First, to minimise the capacitance between the pairs of deflector plates, their connections are brought out on the neck of the tube, thus permitting the use of the latter at higher frequencies. Second, in order to increase the brightness without degrading the deflection sensitivities, another electrode *after* the deflector plates is used to provide additional acceleration to the beam. This is known as *post-deflection acceleration* (PDA).

The deflection of the beam, as measured in millimetres horizontally or vertically on the screen, is directly proportional to the voltage between the corresponding deflector plates; a doubling of deflection indicates a doubling of voltage. It is this property which makes the CRT such a valuable tool for the measurement of very complex waveforms. Fig 3 shows how the CRT is biased.

SPEED

Finally, it is worth mentioning the speeds at which the electrons travel, even in modest systems, such as may be constructed for amateur use. For an EHT of only 600V, the electrons strike the screen at around 1/20 the speed of light, and at 5000V, this becomes 1/7 the speed of light!

How the cathode-ray oscilloscope works

In 'How the Cathode-Ray Tube Works' the principle of the cathode-ray tube (CRT) was discussed. Now it is time to look at a particular piece of equipment which uses the CRT. In the radio amateur's shack, this equipment is usually the cathode-ray oscilloscope (CRO).

BACKGROUND

In that article, the functions of the heater, cathode, grid, anodes, deflector plates and phosphor were explained, together with the operation of the brilliance (or intensity) and focus controls. It is now instructive to see how signals external and internal to the CRO are fed to the CRT, and to understand why the oscilloscope is so useful as a measurement and test tool.

X AND Y

For 90% of its everyday uses, the CRO presents a real-time display of voltage (along the y-axis) against time (along the x-axis). The varying voltage on the y-plates is usually an amplified version of an external signal applied to the y-amplifier through a socket on the front panel (see Fig 1). This amplifier has a large, fixed gain, but is preceded by a wide-range frequency-compensated attenuator in order to accommodate the display of signals of widely varying magnitudes. The calibration scale on the y-attenuator enables the signals to be measured. The signal present on the x-plates varies uniformly with time, and is generated internally by circuits known as the *timebase* and the *x-amplifier*.

THE TIMEBASE

In order to 'draw' the spot at a constant speed across the face of the tube, a sawtooth waveform is used, as shown in Fig 2(a). The waveform consists of two parts: the sweep is a rising voltage which is linear with respect to time and deflects the spot

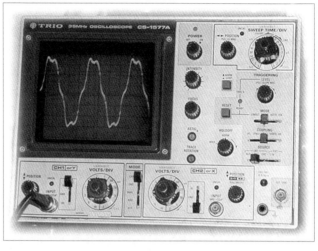

An oscilloscope displaying a complex waveform

Fig 1. Block diagram of the basic cathode-ray oscilloscope

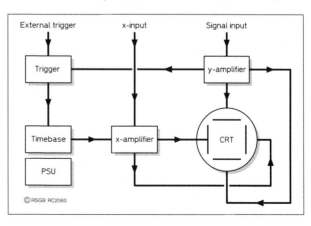

External trigger   x-input   Signal input

Trigger

y-amplifier

Timebase   x-amplifier   CRT

PSU

© RSGB RC2060

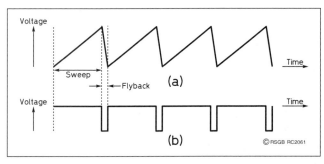

Fig 2. (a) The voltage waveform applied to the x-plates of the CRT. (b) The blanking pulses applied to the CRT grid to suppress the appearance of the flyback on the screen

from left to right across the tube screen; the flyback is the falling section, which should occur very quickly, returning the spot to the left-hand side of the screen.

This process is repeated to give the continuous display of the input waveform. At very high sweep speeds, the flyback may occur in a time comparable with the sweep; when this occurs, the flyback becomes visible and clutters the display. To overcome this problem, *flyback blanking* is used. When the flyback occurs, the negative-going blanking pulse shown in Fig 2(b) is applied to the grid of the CRT; this effectively switches off the beam, and no flyback is visible. Referring to the tube bias circuit given in the previous article, the blanking pulse would be applied to the top end of R1 which is connected directly to the grid of the CRT.

The speed at which the timebase runs is governed by a calibrated rotary switch ('coarse') and an uncalibrated potentiometer ('fine'), both on the front panel. These enable the speed of the timebase to be matched to the frequency of the incoming signal, to display a convenient number of cycles on the screen. The calibration on the timebase control allows the measurement of time intervals to be made from the display, from which signal frequencies can be calculated if required.

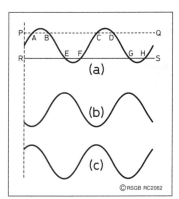

Fig 3. (a) The incoming waveform with two triggering levels shown. (b) The displayed waveform when triggered at points E and G. (c) The displayed waveform when triggered at points A and C

There are three basic modes of timebase operation: *free-running*, where the timebase runs freely at the speed set by the controls; *synchronised*, where the free-running timebase is given 'kicks', derived from the incoming signal, which return it to the left side of the screen and will give a stationary display when the sweep controls are correctly set; *triggered*, where the timebase is inhibited (stopped) and can be triggered to produce a single sweep only, giving a stationary display under all circumstances.

Both the synchronised and triggered modes require level and slope-detecting circuits to operate. These functions are illustrated in Fig 3.

TRIGGERING

An incoming sine wave is shown in Fig 3(a). If the timebase were to be triggered (or synchronised) whenever the voltage level crossed the value indicated by the line RS, it would begin the sweep at two points in every cycle, such as E and F or G and H. If the display is to remain static (the aim of triggering), this is unsuitable. In order to isolate only one point per cycle, the detection of the waveform's slope is also needed. If a negative slope is selected, only points E and G will be found at the level RS, ie one point per cycle. The timebase will begin its sweep at the same point in each cycle, and thus will produce a static display, shown in Fig 3(b).

A positive slope selection would trigger the timebase at F and H. Fig 3(c)

illustrates the display that would be obtained by using a positive slope at level PQ. As Fig 1 shows, it is usual to derive the triggering signal from the incoming signal via the y-amplifier, but most oscilloscopes have a socket for feeding in a trigger signal from an external source. This is most useful when viewing a complex waveform.

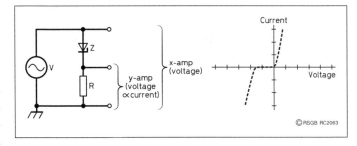

Fig 4. Displaying Zener diode characteristics using a CRO

There is a simple test to check if a timebase is being synchronised or triggered: if the timebase stops when the signal is removed, it is being triggered. Many types of CRO do not distinguish between synchronising and triggering on the front-panel controls.

Finally, what about the 10% of cases where the x-axis is not time but is another signal generated externally? One such application is illustrated in Fig 4.

It uses the y-against-x capability of the CRO to plot the characteristics of a single component, in this case a Zener diode. The voltage signal is fed to the x-amplifier, and generates the x-axis; the voltage developed across R is proportional to the current flowing, and generates the y-axis. In this case, the x-axis scale is 5V per division; the display shows a 10V reverse Zener breakdown, together with the standard forward characteristic of a silicon diode. The voltage source can be derived from a transformer or from a function generator, preferably using a triangular wave. The measurements are not source-waveform-dependent.

Other non-timebase applications include the measurement of frequency using Lissajous figures, but this has little direct application in amateur radio.

SPECIAL MODELS

Oscilloscopes made specifically for amateur radio are usually called *station monitors*; typical examples being the Kenwood SM-220 and SM-230. These are conventional oscilloscopes but have direct HF/VHF access to the y-plates, which allows a radiated RF waveform to be displayed, and trapezoidal displays of HF waveforms to be produced. They also act as panoramic displays with the TS-950 range of transceivers.

Calculating meter shunt values

Panel meters come in a variety of shapes and sizes but relatively few ranges. It would be impractical for a manufacturer to produce a meter for every application. This is not a problem, however, as it is a easy matter to extend the range by connecting a 'shunt' resistor across a meter. This is simply a resistor which carries a known proportion of the total current, leaving the meter to indicate within its own range.

EXAMPLE

Suppose a particular meter (Fig 1(a)) has a resistance of 10Ω and requires a current of 1A for full-scale deflection (FSD). If we connect a resistor of 10Ω (the same resistance as the meter) across it (Fig 1(b)), the current is now shared equally between meter and resistor, so 2A in total will have to flow

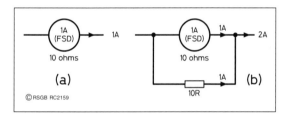

(a)

(b)

©RSGB RC2159

Fig 1. (a) Typical 1A meter might have an internal resistance of 10Ω. (b) To make the meter read 2A FSD, it requires a 10Ω 'shunt'

for 1A to pass through the meter. Thus, the meter indicates full-scale when a total of 2A is flowing through the circuit. A careful distinction must be drawn here between the use of the words 'indicate' and 'read'. In this example, the meter still *reads* 1A, although it is *indicating* that 2A is flowing in total.

If the value of shunt is now halved to 5Ω, 2A will bypass the meter when 1A passes through it. It will therefore indicate 3A at FSD, as shown in Fig 2.

Looking at the junction of the wires to the right of Fig 2, we can see that currents of 2A and 1A are flowing *into* the junction, and a current of 3A is flowing *out of* it. This must always be true because we can't 'lose' current anywhere – what goes into any junction must come out! As an equation, and using the symbol I for current, this becomes:

$$I_{total} = I_{shunt} + I_{meter}$$

Looking at the values of the currents in Fig 2, we can deduce that:

$$\frac{R_{shunt}}{R_{meter}} = \frac{I_{FSD}}{I_{shunt}} \quad \text{or} \quad R_{shunt} = \frac{I_{FSD}}{I_{shunt}} \times R_{meter}$$

where R_{shunt} is the resistance of the shunt and R_{meter} is the resistance of the meter.

By combining the equations for I_{total} and R_{shunt}, we can arrive at a formula that allows us to calculate the value of shunt required to make a meter indicate any current that is larger than its FSD. Bear in mind that you can never make a meter more sensitive; a shunt makes a meter less sensitive, ie it indicates larger currents than its normal full-scale deflection.

$$R_{shunt} = \frac{I_{FSD}}{I_{total} - I_{FSD}} \times R_{meter}$$

I_{total} is the current that we want the meter to indicate, as consideration of Fig 2 will show.

In reality, the value of the shunt resistor is seldom a convenient 'off the shelf' value; it is necessary to make up the value with a series/parallel combination. Alternatively, very low values of resistance can be 'manufactured' by using resistance wire such as 28SWG constantan, which is available from Maplin. This has a resistance of 4.2Ω per metre so, by simple measurement, quite precise resistances can be made. It is best to cut the amount required plus about 3mm at each end for connections; longer lengths will require a former around which the wire can be wound.

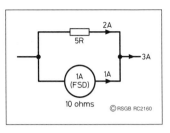

Fig 2. A shunt of half the meter's resistance will take twice as much current

CHOOSING METERS

It is a great help if the actual scale of the meter bears some resemblance to the quantity being measured. For instance, if you want to measure up to 50mA, a 50μA meter would be a good choice. Using the method outline above, try not to exceed a 10:1 ratio between the shunt and meter resistances. If this really *is* necessary (eg to make a 50μA meter indicate 500mA), then there are other ways to do it which are outside the scope of this book.

DISSIPATING POWER

Remember that all resistors dissipate power, and shunt resistors often need to dissipate quite a lot. For example, take the case illustrated in Fig 2. A 5Ω shunt is passing 2A. The formula for power in a resistor is:

$$P = I^2 \times R$$

Putting in the values gives:

$$P = 2^2 \times 5 = 20W$$

which, almost certainly, would require a shunt made from a length of constantan wire wound, if possible, on a ceramic former and located where a free flow of air would help to dissipate the heat produced.

Transformers

Transformers are widely used in all branches of electronics. One of their well-known uses is in power applications, where they are used to 'transform' the operating voltage from one value to another, hence the name. They also serve to isolate the circuit at the output from a direct connection to the primary circuit. In this way they transfer power from one circuit to another with no direct connection. Very large transformers are used on the National Grid to change the line voltages between the different values required. However, for the radio amateur or home enthusiast, transformers are commonly seen in power supplies. Transformers are also widely used in other circuits from audio up to radio frequencies, where their properties are employed to couple different stages within the equipment.

WHAT IS A TRANSFORMER?

A basic transformer consists of two windings, and the circuit symbol is shown in Fig 1. The windings are known as the *primary* and the *secondary*. In essence, power enters the primary and leaves the secondary. Some transformers have more windings but the principle of operation is still the same.

There are two main principles used in a transformer, relating to the currents in the windings and magnetic fields that they produce.

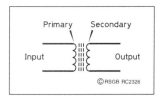

Fig 1. Circuit symbol of a transformer

First, it is found that a current flowing in a wire sets up a magnetic field around it. The magnitude of this field is proportional to the current flowing in the wire. It is also found that if the wire is wound into a coil then the intensity of the magnetic field inside the coil is greater than that around the single wire. If this electrically generated magnetic field is placed in an existing magnetic field then a force will be exerted on the wire carrying the current, in the same way that two magnets placed close to one another will either attract or repel one another. It is this phenomenon that is used in electric motors, meters, and a number of other electric systems.

Second, if a magnetic field around a conductor changes, then an electric current will be induced in the conductor. One example of this can occur if a magnet is moved close to a wire or a coil. Under these circumstances an electric current will be induced but only while the magnet is moving.

The combination of the two effects occurs when two wires or two coils are placed close together. When a current changes in the first coil, this results in a change in the magnetic flux (field) and in turn results in a current being induced in the second coil. This is the basic concept behind a transformer, and from this it can be seen that it will only operate when a changing or alternating current is passing through the input or primary circuit.

For a current to flow in the transformer secondary, an *electromotive force* (EMF) must be present. This potential difference or voltage at the output is dependent upon the ratio of turns in the transformer. If more turns are present on the primary than on the secondary, the voltage at the output will be less than that at the input. Conversely, if there are more turns on the secondary than on the primary, the voltage appearing across the secondary will be greater than that across the primary. In fact, the voltage can easily be calculated from a knowledge of the turns ratio:

$$\frac{V_S}{V_P} = \frac{N_S}{N_P}$$

where V_S is the secondary voltage, V_P is the primary voltage, N_P is the number of turns on the primary, and N_S is the number of turns on the secondary. The ratio N_S/N_P is known as the *turns ratio* of the transformer.

Fig 2 shows the basic parameters associated with a transformer. If the turns ratio N_S/N_P is greater than unity, the transformer will produce a higher voltage at the output than is present at the input, and is said to be a *step-up transformer*. Similarly, a transformer with a turns ratio of less than unity is a *step-down transformer*. When the turns ratio is unity, the output and input voltages are the same, and the transformer is used simply to isolate the two circuits; there is no voltage change but there is no electrical connection between primary and secondary. This is known as an *isolating transformer*.

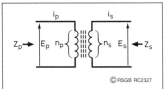

Fig 2. Parameters associated with a transformer

There are several other factors that can be easily calculated. The first is the ratio of input and output currents and voltages. Ignoring any losses (and the losses in mains transformers are usually very small and therefore ignored), the output power is equal to the input power. From this it is possible – using the simple formula shown below – to calculate a voltage or current if the three other values are known. Remember that power is the product of voltage and current.

$$V_P \times I_P = V_S \times I_S$$

For example, take the case of a mains transformer that delivers 25V at 1A. With an input voltage of 250V, this means that the input current is 1/10A.

For isolation transformers, the current and voltage at the output will be the same as that at the input. However, where the turns ratio is not 1:1, the voltage and current ratio will be different at the input and the output. From the simple relationship shown above it will be seen that the current ratio is the inverse of the voltage ratio:

$$\frac{I_S}{I_P} = \frac{V_P}{V_S}$$

For example, a transformer with a turns ratio of 2:1 may have a 20V input with a current of 1A, whereas at the output the voltage will be 10V at 2A (voltage ratio = 10/20 = 1/2; current ratio = 2/1 = 2).

IMPEDANCE

As the ratio of voltage to current determines the impedance, it can be seen that the transformer changes the impedance between the input and the output. In fact the impedance varies as the square of the turns ratio:

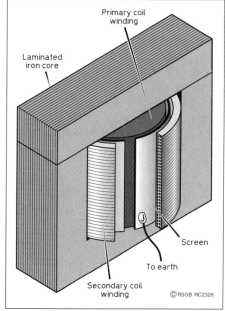

$$\frac{Z_P}{Z_S} = \frac{N_P^2}{N_S^2}$$

IN USE

Transformers are widely used in many applications in radio and electronics. One of their main applications is within mains power supplies. Here the transformer is used to change the incoming mains voltage (around 240V in the UK) to the required voltage to supply the equipment.

With most of today's equipment using semiconductor technology, the voltages that are required are much lower than the incoming mains. In addition to this, the transformer isolates the supply on the secondary from the mains, thereby making the secondary supply much safer. If the supply were taken directly from the mains there would be a much greater risk of electric shock.

A transformer used in a power supply is generally wound on an iron core. This is used to concentrate the magnetic field and ensures that the 'coupling' between the primary and secondary is very tight. In this way, the efficiency is kept as high as possible. However, it is very important to ensure that this core does not act as a single-turn winding. To prevent this happening, the sections of the core are insulated from one another as shown in Fig 3. In fact the core is made up from several plates, each interleaved but insulated from one another. Also, the two windings of a power transformer are well insulated from one another to prevent any likelihood of the secondary winding becoming live.

Although one of the major uses for the transformer is for transforming supply or mains voltages to a new level, it also has a variety of other applications. When valves were widely used, they were employed in audio applications to enable low-impedance loudspeakers to be driven by valve circuits that had a relatively high output impedance. They are also used for radio-frequency applications.

The fact that they can simultaneously:

- isolate the direct current components of the signal;
- act as impedance transformers;
- act as tuned circuits;

means that they are a vital element in many circuits. Fig 4 shows a circuit where a transformer is used as an interstage element in a radio intermediate

Top: **A typical mains transformer, as might be used in a low-voltage power supply**

Above: **Fig 3. Physical construction of a mains transformer**

Primary coil winding

Laminated iron core

Screen

To earth

Secondary coil winding

©RSGB RC2328

frequency amplifier. In many portable receivers, these IF transformers provide the selectivity for the receiver.

In the example shown, it can be seen that the primary of the transformer is tuned using a capacitor to bring it to resonance. Adjustment of the resonant frequency is normally made using a core that can be screwed in and out to vary the inductance of the coil. The transformer matches the higher impedance of the collector stage of the previous stage to the lower impedance of the following stage, and also serves to isolate the different steady-state voltages on the collector of the previous stage from the base of the following stage. If the two circuits were not isolated from one another, the DC bias conditions for both transistors would be disturbed and neither stage would operate correctly. By using a transformer, the stages can be connected for AC signals while still maintaining the DC bias conditions.

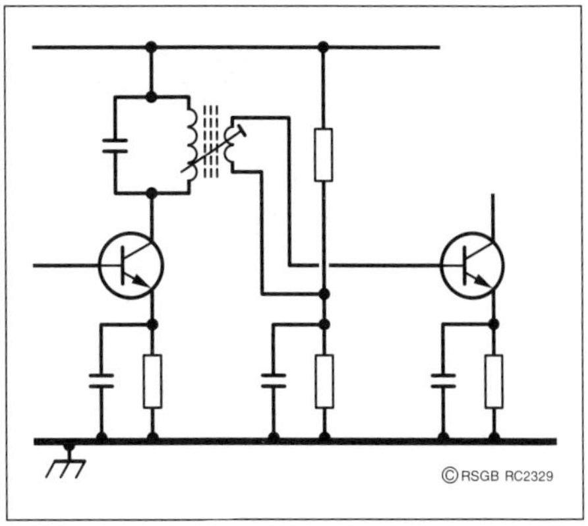

Fig 4. A transformer used as inter-stage coupling in the intermediate frequency stage of a radio receiver

SUMMARY

The transformer is an invaluable component in today's electronics scene. Despite the fact that integrated circuits and other semiconductor devices seem to be used in ever-increasing quantities, there is no substitute for the transformer. The fact that it is able to isolate and transfer power from one circuit to another, while changing the impedance, ensures that it is uniquely placed as a tool for electronics designers.

Voltage regulation

Nothing is perfect; the AC mains, transformers, resistors, transistors, integrated circuits – we know how they *ought* to behave but how they *really* behave can be rather different.

THINGS CHANGE . . .

The AC mains is not constant, particularly if you live out in the country and are served by several miles of overhead cable. Your next-door neighbour switches on her electric oven, and your lights dim! Transformers have losses in the resistance of their windings, and these losses vary depending on the load at any given time. Resistors produce a voltage drop when a current flows through them. The current may cause a resistor to heat up unduly, change its resistance, and produce a different voltage drop. Transistors are also sensitive to changes in temperature, and may cause major problems in a poorly designed circuit.

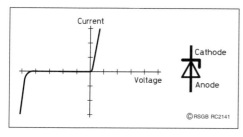

Suppose you are building a circuit using TTL logic chips. On the data sheet specifying how they should be used, you see that the TTL family of integrated circuits requires a supply of 5.00 ± 0.25V. Whatever happens in the rest of your equipment, and as a result of any mains variation, must not cause more than 0.25V variation of the 5V supply. We shall now see how we can stabilise our voltage supplies so that they are virtually independent of input voltage variations.

Fig 1. The circuit symbol and characteristic of a typical Zener diode

THE ZENER DIODE

For voltages below about 30V, the heart of almost all voltage stabilisation (or regulation) circuits is the Zener diode. This is a discrete silicon device, which has the normal diode characteristic when forward-biased (ie when the anode is positive with respect to the cathode), but when reverse-biased it passes negligible current up to a well-defined point at which the current increases sharply. This is illustrated in Fig 1, which also shows the circuit symbol.

The point at which the reverse current starts to flow can be closely controlled in manufacture between about 3V to 200V. If such a Zener diode is intentionally biased into this steep region, the voltage across the diode varies very little for large changes in current; the device is serving to stabilise (or regulate) the voltage across it. A typical circuit is shown in Fig 2. The value of R is given by the equation

$$R = \frac{V_{\text{out}}(V_{\text{in}} - V_{\text{out}})}{P}$$

where V_{in} is the unstabilised input voltage, V_{out} is the stabilised output voltage (the Zener voltage), and P is the Zener dissipation in watts (see text).

Zener diodes are specified by two parameters – the Zener voltage and the maximum power dissipation of the device. For example, all Zener diodes of the types BZY88C and BZX55C have 0.5W dissipation and are by far the commonest. A 15V Zener would be marked BZY88C15V; you are expected to know it is a 0.5W device!

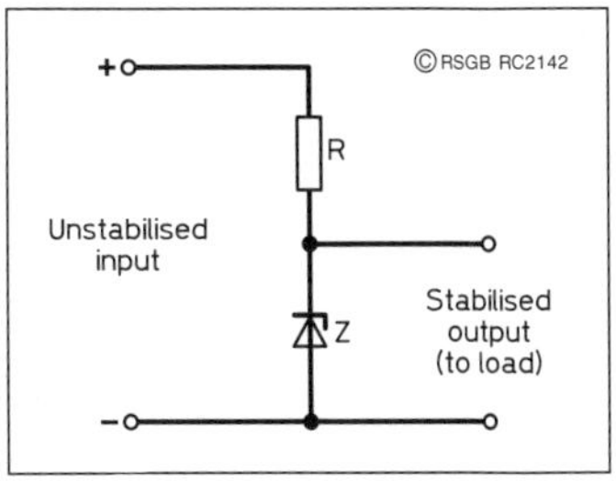

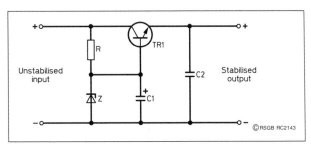

Fig 2. A simple circuit that produces a stabilised output from an unstabilised input

SEE-SAW ACTION

A see-saw is a simplistic, but very useful, analogy of how the Zener works in a circuit. In Fig 2, the current from the unstabilised supply has two paths – through R and Z to earth, or through R and the load connected across Z. Suppose we have chosen a 10V Zener of the BZY88C series. It is a 0.5W device and therefore it will draw a current I given by:

$$I = \frac{P}{V} = \frac{0.5}{10} = 0.05\text{A or } 50\text{mA}$$

Suppose no load is connected. All 50mA will go through the Zener as there is no other route. Suppose the load now takes 5mA; here the current seesaw comes in. With 5mA going through the load, 45mA now goes through the Zener. If the load takes 10mA, then 40mA flows through the Zener. This keeps the total current constant (from the unstabilised supply), and thus helps to keep the voltages constant also.

This see-saw action is not perfect, particularly if the load takes more and more current. As the current through the Zener approaches zero, the seesaw action fails and the stabilisation becomes inoperative. This circuit is simple and works well, provided the demands of the load are small, but it has one big disadvantage – your power supply is delivering the maximum current at all times, whether or not the load takes any. Notice in the example that the Zener takes 50mA when there is no load. Notice also that the unstabilised voltage must always be significantly greater than the stabilised voltage because of the voltage drop across R. It is inadvisable to try to construct a 12V stabilised supply from a 13V unstabilised input! As the range of currents demanded by the load increases, so also should the difference between the supply voltage and the regulated voltage.

A REFERENCE VOLTAGE

Removing the need for the power supply to supply the maximum power at all times can be achieved by keeping the current through the Zener constant. In this way, it does not need to dissipate maximum power and the regulation it provides is much better. It is used more as a reference voltage than as a type of voltage source. This idea is illustrated in Fig 3.

Fig 3. An improved circuit

The stabilised voltage produced by the Zener diode is applied to the base of the transistor, which acts as an emitter follower, producing about 0.6V less than the voltage across the Zener but at a current limited only by the

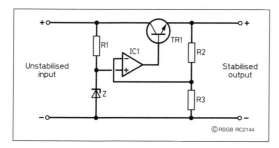

Fig 4. Using an operational amplifier to improve the circuit

capacity of the power supply and the characteristics of the series or pass transistor TR1 which usually requires a heatsink. There is a shunt-stabilising version of this circuit but it is little used because of its inefficiency.

To take this circuit to its logical conclusion and produce a well-stabilised supply, the circuit of Fig 4 is common. This produces, by means of the integrated circuit op-amp, a very sensitive comparison between the Zener voltage, V_Z, and the output voltage divided down by R2 and R3. This allows the output voltage to be greater than the Zener voltage and, depending upon the voltages and currents involved, can obviate the need for extra smoothing capacitors. Readers interested in refinements such as foldback current limiting and crowbar protection are referred to references [1] and [2]. Needless to say, integrated circuits have been produced that possess all the qualities (and some refinements, too) discussed so far but for the lower output current ranges only. These are typified by fixed positive-voltage devices such as the LM78**CT series of 1A regulators and their negative-voltage equivalents, the LM79**CT series. The LM338K is a 5A regulator for output voltages between 1.2V and 32V. The LM317T is a popular 1.5A variable regulator.

HIGHER VOLTAGES

When stabilisation is required at high voltages, solid-state regulation gives way to the regulation afforded by ionised gases at very low pressures. The use of neon and hydrogen discharge devices is beyond the scope of this article, but basic details may be found in reference [3].

REFERENCES

[1] *ARRL Handbook*, 1998 edn, p11.15.
[2] *Radio Communication Handbook*, 6th edn, RSGB, 1994, p3.13.
[3] *The Services' Textbook of Radio*, Vol 3, HMSO, 1963, pp180–198.

Speech processing

There is nothing like a demonstration to illustrate a point, so I would strongly recommend this experiment. On the FM waveband, tune in BBC Radio 4 and adjust the volume to give a comfortable listening level on ordinary speech. Then, retune to a commercial 'pop' station. Whether the content is music or speech, you will probably reach for the control to turn down the volume. The reason for this is the subject of this article, although it is its application to amateur radio that concerns us most. Just like us (but even more so), the broadcasting companies must not over-modulate or over-deviate, so why is speech on one broadcast so much louder than on another?

FIRST STEPS

The waveform (as might be shown on a cathode ray oscilloscope) of male speech, taken from a discussion programme on BBC Radio 4, is shown in Fig 1(a). Fig 1(b) shows the waveform of male speech from a commercial radio station, and Fig 1(c) the waveform of music taken from the same station. All three recordings were made with the same audio gain setting (with no audio AGC), and are of the same length, so they are thus directly comparable. The equipment used was an Icom IC-PCR1000 receiver, a Soundblaster® PC sound card, and proprietary software for recording wave files. To analyse these waveforms quickly, I wrote a short program in Visual Basic 6.0, which analyses the waveform and calculates the power (in arbitrary units) in each wave. The results are summarised thus:

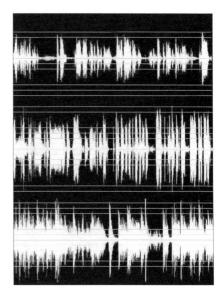

Fig 1. (a) A section of male speech from BBC Radio 4. (b) A section of male speech from a commercial radio station. (c) Vocal pop music from the same commercial radio station

Radio 4 speech: Power = 138
Commercial radio speech: Power = 504
Commercial radio music: Power = 522

For these random selections of programme output, you can see that there is almost four times as much power in commercial radio speech as there is in Radio 4 speech, and that the music played on commercial radio is only marginally louder than the DJ's pearls of wisdom! This process of 'making everything louder' is called *dynamic range compression, contrast compression* or, simply, *compression*. As we amateurs are concerned only with speech, we know it as *speech compression* or *speech processing* and, for reasons to be discussed later, we use it only when working on SSB.

THE PROBLEM

Speech is remarkably 'spiky' in nature, as Figs 1(a) and 1(b) show when viewed on a cathode-ray oscilloscope. This spikiness causes us problems when we have to set the modulation level on our transmitters. On sideband transmissions, over-modulation causes splatter at the very least, and on FM transmissions it produces over-deviation.

To avoid this, we must set the transmitter drive so that the *peaks* of our audio waveform do not over-modulate. To illustrate this, use Fig 1(a), and lay a ruler on it horizontally so that it *just* touches the biggest negative-going peak (we choose a negative-going rather than a positive-going peak, so that the ruler does not hide the waveform). You will now notice one thing – that the average modulating voltage (compared with our self-imposed maximum) is very low. In practical terms, our transmission will lack 'punch', and will not be heard very well under difficult conditions.

SOME SOLUTIONS

In very basic terms, what we need to do is to turn up the audio gain when the speech is quiet and turn it down when the speech is loud. Unfortunately, these variations of loudness occur very quickly, often between syllables, and the technique is something which cannot be done manually.

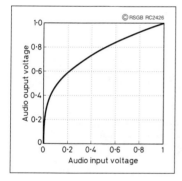

Fig 2. Using a non-linear transfer characteristic as the basis of a speech processor

Two basic methods are used, one of which involves using an amplifier of non-linear transfer characteristic; the other uses automatic control of the audio gain. It should be understood at the outset that, whatever technique is used, it *distorts* the speech waveform. This distortion is (hopefully) controlled so as to improve the intelligibility, even if the result is sometimes unnatural.

Using a non-linear transfer characteristic is, in theory, probably the most attractive solution. A transfer characteristic is the graphical representation of the output voltage from a device compared with the input voltage. Normally this should be a straight line, indicating that the output and input voltages are directly proportional.

To achieve the desired compression, look at the characteristic of Fig 2; notice that, for example, a small input of 0.1V will produce an output of about 0.5V (a gain of about 5), 0.2V will produce an output of about 0.6V (a gain of about 3), whereas a large input of 1V will produce an output of 1V (a gain of unity). Provided the input voltage never exceeds 1V, the circuit will produce an increasing gain at lower voltages. The main problem with a technique such as this is that a circuit to implement it is quite difficult to design. It does have the advantage (see later) that nothing *physically* changes as a result of changing input levels.

The other techniques are simpler to implement, but have their own disadvantages. Manually altering the audio gain control may be completely impractical but circuits can be made which will achieve the same result automatically.

The operations involved are straightforward:

(a) Rectify the audio voltage. The average value of an audio signal is zero, so rectification is needed to 'lop off' the lower half of the signal so that it *does* have a mean value.

(b) Feed the rectified signal into a low-pass filter, a simple connection of one resistor and one capacitor (see Fig 3). The output from this filter is the mean value of the signal over short time periods, the audio components having been short-circuited by the capacitor.

(c) This varying mean value is then used to control the gain of another audio amplifier, the gain being reduced as the mean value increases. A block diagram of the circuit is shown in Fig 3. A buffer amplifier is a device which helps to separate the rectifier and filter from preceding circuits and to supply enough gain to operate the rectifier cleanly.

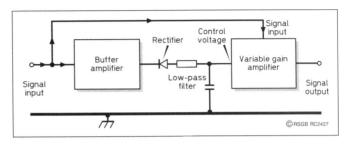

A simple gain control circuit, eminently suitable as the basis for experimental work, is shown in Fig 4. An n-channel JFET is used as one of a pair of feedback resistors in a standard non-inverting op-amp circuit. Connected this way, the resistance between the JFET drain and source depends upon the voltage applied to its gate. The overall gain [1] is adjustable between 1 and 1000 by changing the voltage at the control input.

Fig 3. Block diagram of a simple audio AGC circuit

ATTACK AND DECAY

While Fig 4 is quite a simple and acceptable circuit to use for automatic audio gain control, the derivation of the control voltage to operate it can be quite tricky. The simplest circuit, comprising the rectifier, capacitor and resistor shown in Fig 3, is a good starting point. The values of the resistor and capacitor need to be varied in order that the circuit will act quickly to reduce the gain when a loud sound suddenly appears (this is known as 'attack'), yet will take a little longer to return to its initial gain (known as 'decay') in case another loud sound follows on quickly. Getting the attack and decay right is quite an art, and a circuit more complex than that of Fig 3 is usually needed.

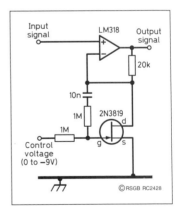

THE PROBLEMS

No circuit like this is without its problems. Correct attack and decay are obvious candidates, and the solution is somewhat subjective, because no two people speak in exactly the same way. However, the overriding problem with all AGC-based speech processing circuits is that a loud sound *must* get through to the output *before* the circuit can react to reduce the gain. This means that, in unfortunate cases, transient peaks which are louder than they ought to be will get through to the output and could still cause transmitter over-modulation.

The overall drive must be backed off a little to allow for this. Nevertheless, speech processing, when properly used, is of great benefit to the SSB station working in crowded band conditions.

Fig 4. A simple amplifier circuit where the gain is dependent upon the control voltage. The circuit supply voltage should be around ±15V

203

In professional circuits using this principle, the signal between the main input and the input to the gain-controlled amplifier in Fig 3 is subject to a short delay; which means that the gain *can* be turned down *just in time* for the delayed signal which caused it to reach the amplifier!

OTHER METHODS

Speech processing may also be achieved (and more effectively so) by performing the control operation on the RF signal, or by using DSP. Both these techniques are beyond the scope of this article, however.

FINALLY

We have not discussed speech processing in the context of FM. FM is essentially a short-distance mode, and interference from adjacent stations is not a problem. Saying this removes the need for processing, hence the restriction of the discussions to SSB.

If circuits are useful, you can bet that they have been made into integrated circuits! The speech processor is no exception, and one such device is the VOGAD in an 8-pin DIL chip, the SL6270 (Maplin order code UM73Q).

REFERENCE

[1] *The Art of Electronics*, Horowitz and Hill, Cambridge University Press, 1988, p240/41.

Safety, operating practice and the law

Just about everyone has heard of the Health and Safety at Work etc Act 1974 (HSAW) and quite a few will have heard of the Factories Act 1961, and the Offices, Shops, and Railway Premises Act 1963. The latter two, and indeed virtually all health and safety legislation, are made under the HSAW Act. Rather fewer, however, will have come across the Electricity at Work Regulations 1989. The purpose of this article is not to shock the living daylights out of those of you who enjoy the opportunity to put on display stations or those who provide emergency communications, but to ensure that you know what legally is expected of you, what you should do to comply, and what might happen if you don't and it all goes wrong.

AT HOME

First, however, let's consider your radio installation at home. Within limits, you can do almost what you want, as far as health and safety is concerned. The HSAW Act does not extend to the individual at home as far as a hobby is concerned. But a word of caution! If your radio installation causes death, injury, or loss to persons on your property – even if they are there without your permission – and it can be shown that your installation is well below the standard expected by the reasonable man on the street, there is the possibility that your installation could be viewed as a 'man trap'. Setting or installing any type of man trap is illegal under the Offences Against the Person Act and other legislation. An example might be, for instance, biasing your HF long wire with 500V to stop anyone attempting to steal it!

Every year, many radio amateurs put on demonstration stations. These may be at fêtes, craft exhibitions, schools, clubs, and so forth. I am sorry to say that of the dozens I have visited over the years, at more than half of them I have seen practices that range from the "could have been better" right up to the "I'll order the coffin now" scenario. In general terms, the problems stem from two areas: first, electrical safety: second, antenna system erection, positioning and taking down. In reality, I suppose the problems stem from poor organisation and, although it is not a defence, ignorance.

ELECTRICAL SAFETY

We all know that electricity can kill. But do you know how much current you need and for how long it needs to be flowing? "But it's all right," I hear you say, "there's a fuse in the plug!" Well, unfortunately, the fuse will offer the individual very little, if any, protection from electrocution. The fuse is there to protect the wiring between it and the supply, not the appliance or the user. The requirements for a circuit protected by a standard BS1362 fuse (the type fitted in a 13A plug), immaterial of the rating of the fuse, is that the fuse will disconnect the current within a maximum time of 0.4s (regulation

A car battery can cause an explosion, burns and a fire

413-02-08 of the 16th Edition of the *IEE Wiring Regulations*). For this to occur, the maximum earth loop impedance must not be greater than 2.3Ω. From Ohm's law, for a 240V supply, this would equate to a current of just over 100A, and if the impedance is considerably lower, say 0.024Ω, as it may be if you are close to the distribution grid transformer, you are talking about currents of 10,000A.

These is not a fantasy value, it is the magnitudes of current that can flow in fault conditions before the fuse ruptures, and is generally called the *prospective short-circuit current* (PSCC).

Most of us are aware of the residual current device (RCD) – *not* to be confused with the earth leakage circuit breaker (ELCB), which is something entirely different and no longer permitted in domestic installations. The RCD parameter used for 'people protection' is that the device will operate within 30ms – many magnitudes faster than a fuse. It should be noted that the RCD is looking for an imbalance of current between the live and neutral and hence, if you become connected between the live and neutral, you will present a balanced load and the RCD will not operate. Usually, however, there is a fortuitous path to earth and the RCD operates.

POTENTIAL PROBLEMS

The usual electrical problems associated with demonstration stations are:

* Excessively long extension leads
* Plugs and sockets of an inappropriate type for outdoor use
* No, or inadequate, mechanical protection for extension leads
* No RCD at the start of the supply point
* Cables too small for the size of fuse
* Use of adapters instead of proper plug boards
* Taped joints, terminal blocks, or match sticks!
* Access to plugs and sockets by the public
* Poor or no earths
* No appropriate testing of installation before use

"But our station runs off 12V car batteries, so it presents no electric shock risk". Maybe. But your car battery *can* produce several hundred amps and a short circuit can cause an explosion, burns or a fire.

Let's take a quick look at some of the problems listed above:

* Long extension leads of low cross-sectional area. Problem: high loop impedance – the fuse may take a long time (greater than 400ms) to rupture or may not blow at all.

- Cable protection – mechanical protection is needed to prevent damage by vehicles, shoes or animals.
- Plugs and sockets – should be to BS4343 if used outside.
- RCD – should be at the start of the cable run to protect against all possibilities.
- Adapters – usually provide poor connections, particularly the earth pin. Use a plug board.
- Connections – if you haven't got the correct connector, stop! Taped joints, terminal blocks, and similar 'quick fixes' equate to 'quick death'. The public, and in particular children, must not be able to access plugs, sockets, or switchgear.
- Earthing – this must be adequate for the purpose. A meat skewer of 7/.22 is of no use whatsoever! You need a 1m earth rod and 10mm² cable, and the rod must be 900mm in the ground.
- Testing – check that the safety system works. Are L and N the correct way round? Is the ground impedance low enough? Is the installation out of harm's way?

ANTENNA SAFETY

"How on earth can an antenna be unsafe?" Whilst 'on earth' it may be much safer than it is on the top of a 20m tower, but even when it is lying on the ground you must be aware of trip hazards. A crossed Yagi on the floor presents something akin to the defences used in the Second World War to stop tanks. It can easily stop a human being.

Antenna safety covers a number of functions and conditions, including erection, use, and dismantling. Each of these activities brings with it a range of hazards.

A primary operation of any antenna erection is to survey the proposed site well ahead of the day. You should be looking for the proximity of other overhead structures and, in particular, cables. Due regard needs to be given to the route the feeder will take to the station. Does it need to be given protection from vehicles and pedestrians? If open-wire feeders are used they must be well out of reach. Guy wires should be visible (ie marked with hazard-marking tape) or fenced off. Stakes should be similarly marked, eg with traffic cones. Feeder wires and guy wires can easily garotte someone! The antenna system itself needs to be properly attached to the mast. Bailing twine and insulation tape are not appropriate! High-quality clamps in good condition and tightened with proper spanners represent the only satisfactory method.

Erecting the mast/tower should be practised well before the day. Written instructions should be available and followed. There should be nobody in the area without a reason to be there. All those involved must wear hard hats and appropriate footwear. One person should be declared in charge and (s)he is the only one who should be giving instructions. You need to consider what will happen if it all goes wrong. Is there plenty of clearance if the mast falls over? Is it immune from the attentions of children? Are warning signs posted? Have you got an adequate first-aid kit immediately to hand? And finally, make sure you have permission to erect the antenna where you intend to put it. During use, a periodic inspection of the antenna and mast should take place. This should alert you to anything which has become loose either on its own or due to unwelcome visitors. It also gives you a chance to ensure that someone else hasn't tied guy ropes, bunting, or floodlights to part of *your* antenna system.

Dismantling the antenna system is usually the reverse process of putting it up. But, here again, many of the problems involved with the erection are present and the same precautions need to be taken.

WHAT IF IT ALL GOES WRONG?

If an accident occurs, the initial reaction is often to rush in and see if you can help. Whilst the intent is laudable, the practice isn't. The first step you should take is to ensure that *you* are not in any danger. It's no use grabbing hold of someone who is in contact with the mains unless you want to join them, or trying to remove someone from under a collapsed mast when the rest of the antenna system is going to fall on top of you. Make the incident site safe. Give whatever first-aid is appropriate. Keep people away unless they have a reason to be there.

Call for help from the emergency services, inform the event organiser and, depending on the nature of the event, get the organiser to contact the Health and Safety Executive (HSE). If you have a camera, or better still a video recorder, record the salient parts of the incident site. Make notes of what happened and don't let anyone interfere with the site or take home 'souvenirs'. If possible get the personal details of the injured party and pass to the ambulance crew and the police. Do not release these details to anyone else. It is the duty of the police to advise relatives of the accident.

LEGAL ACTION

If the HSE and/or police are involved, it is their decision if a prosecution is warranted. As far as the HSE is concerned, it will be looking for breaches of the legislation mentioned at the beginning of this article. The police would generally be interested if someone had caused criminal damage, eg cut through the guy ropes which had led to the accident. This in turn could lead to further charges including manslaughter.

Depending on the seriousness of the offence, it can either be tried at the Magistrates' Court or referred to the Crown Court. For breaches of H&S legislation, the Magistrates' Court can fine up to £25,000. The Crown Court has unlimited powers, including imprisonment. Of course, things may not end there. The injured party (or relatives in the event of a death) may decide to pursue a case in the civil court for substantial damages. In all cases, the defence will be to show and prove that all reasonable due care had been taken and diligence shown, that the event had been thought through and planned, and that consideration had been given as to what action to take if it went wrong and appropriate plans made.

It is worth remembering that you cannot legislate for fools or the igno-rant. The best tack here is to take out an insurance policy – but then you did that in the initial planning stage, didn't you?

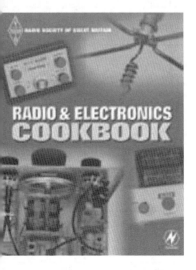

RSGB ORDER FORM

ORDERED BY

ORDER NO.   **DATE**

DELIVER TO

Code	Description	Price	Qty	Total
1-872309-50-0	Amateur Radio ~ the first 100 years	£49.99		
1-872309-70-4	Amateur Radio Explained	£9.99		
1-872309-63-1	Amateur Radio Operating Manual	£24.99		
1-872309-36-4	Antenna Experimenter's Guide	£17.99		
1-872309-72-0	Antenna File	£18.99		
0-7506-4947-X	Antenna Toolkit 2nd Edition	£24.99		
1-872309-89-5	Antenna Topics	£18.99		
1-872309-54-2	Backyard Antennas	£18.99		
1-872309-87-9	Deluxe Log book	£4.99		
1-872309-82-8	Digital Modes for all Occasions	£16.99		
1-872309-80-1	Foundation Licence Now!	£3.99		
1-872309-00-3	G-QRP Circuit Handbook	£9.99		
1-872309-48-8	Guide to EMC	£19.99		
1-872309-58-5	Guide to VHF/UHF Amateur Radio	£8.99		
1-872309-75-5	HF Amateur Radio	£13.99		
1-872309-08-9	HF Antenna Collection	£9.99		
1-872309-15-1	HF Antennas for all Locations	£12.99		
1-872309-86-0	Intermediate Licence Handbook	£5.79		
1-872309-83-6	International Microwave Handbook	£24.99		
1-872309-81-X	IOTA Directory	£9.99		
1-872309-65-8	Low Frequency Experimenter's Handbook	£18.99		
1-872309-73-9	Low Power Scrapbook	£12.99		
1-872309-12-7	Microwave Handbook "Bands & Equipment"	£18.99		
1-872309-77-1	Mobile Radio Handbook	£13.99		
1-872309-26-7	Morse Code for Radio Amateurs	£4.99		
1-872309-31-3	Packet Radio Primer	£9.99		
1-872309-40-2	PMR Conversion Handbook	£16.99		
1-872309-88-7	Practical Projects	£13.99		
1-872309-85-2	Prefix Guide	£8.99		
0-705652-1-44	Radio & Electronics Cookbook	£16.99		
1-872309-45-3	Radio Amateur's Examination Manual	£14.99		
1-872309-53-8	Radio Communication Handbook	£29.99		
1-872309-30-5	Radio Data Reference Book	£14.99		
1-872309-18-6	RAE Revision Notes	£5.00		
1-872309-79-8	Receiving Log book	£4.99		
0-7506-4844-9	RF Components & Design	£22.50		
1-872309-84-4	RSGB Yearbook	£15.99		
1-872309-71-2	Technical Compendium	£17.99		
1-872309-20-8	Technical Topics Scrapbook 1985-89	£9.99		
1-872309-51-8	Technical Topics Scrapbook 1990-94	£13.99		
1-872309-61-3	Technical Topics Scrapbook 1995-99	£14.99		
1-872309-23-2	Test Equipment for the Radio Amateur	£12.99		
1-872309-78-X	Value Log book	£4.99		
1-872309-76-3	VHF/UHF Antennas	£13.99		
1-872309-42-9	VHF/UHF Handbook	£19.99		
0-900612-09-6	World at Their Fingertips	£9.99		
1-872309-43-7	Your First Amateur Station	£7.99		
1-872309-38-0	Your First Packet Station	£7.99		
1-872309-49-6	Your Guide to Propagation	£9.99		

Post & Packing		P & P		
UK only - £1.50 for 1 item £2.95 for 2 or more items		Discount		
Rest of World - £2.00 for 1 item 4.00 for 2 & £0.50 for each extra item		Total		

RSGB, Lambda House, Cranborne Road, Potters Bar, Herts EN6 3JE UK

Tel: 0870 904 7373 Fax: 0870 904 7374 E-mail sales@rsgb.org.uk   *E&OE*